A Genealogy of Appetite in the Sexual Sciences

Jacinthe Flore

A Genealogy of Appetite in the Sexual Sciences

palgrave
macmillan

Jacinthe Flore
Royal Melbourne Institute of Technology
Melbourne, VIC, Australia

ISBN 978-3-030-39422-6 ISBN 978-3-030-39423-3 (eBook)
https://doi.org/10.1007/978-3-030-39423-3

Cover illustration: Chronicle / Alamy Stock Photo
Cover design by eStudio Calamar

This Palgrave Macmillan imprint is published by the registered company Springer Nature Switzerland AG.
The registered company address is: Gewerbestrasse 11, 6330 Cham, Switzerland

For Marc Trabsky

Acknowledgments

The trajectory of this book is marked by a series of departures and arrivals. I am thankful to the individuals I have encountered along this journey, who have provided intellectual debate, provocative discourse and friendship. The book began as a doctoral thesis at La Trobe University, which was supported by a La Trobe University Research Scholarship. My interest in writing a history of sexuality, however, was sparked by Farhad Khoyratty and Naseem Lallmahomed-Aumeerally during my Bachelor of Arts at the University of Mauritius. I would like to express my sincere gratitude for the mentorship and inspiring critique of Steven Angelides and Carolyn D'Cruz, and the support of Gary Dowsett and Wendy Mee in the early stages of this project, as well as the collegiality of doctoral students at the Australian Research Centre for Sex, Health and Society and the Gender, Sexuality and Diversity Studies program, La Trobe University.

I am grateful for the encouragement of my colleagues and friends in the Social and Global Studies Centre at RMIT University for finishing the manuscript. For engaging conversations and valuable feedback on the project, I thank Lisa Downing, Donna Drucker, Jennifer Germon, Lesley Hall, Anne-Marie Jutel, Tania Lewis, Fran Martin, Kiran Pienaar, Jordy Silverstein, Juliet Watson and Jeffrey Weeks.

I wish to thank the staff at the *Bibliothèque nationale de France*, the *Collège de France* and Wellcome Library in the United Kingdom for their hospitality and support with accessing their archives. Archival research in England and France in 2013 was generously funded by a Humanities and Social Sciences Research Grant from La Trobe University. I am indebted to the invaluable support provided by Palgrave Macmillan in completing this book, particularly Emily Russell.

My family in Mauritius, South Africa and Australia has been central to this work. I wish to express my deepest gratitude to my late mother Anne-Marie Flore (*née* Dupuche), whose love and support I could never fully capture in words; she has been essential to all my scholarly pursuits and the memory of her teaching me to read and write is forever etched in my mind. For their encouragement, love and kindness, despite the distance, I thank my father Hervé, my siblings Joëlle, Vincent and Fabien, my niece Mélanie, and my extended family. I am grateful to Paméla Marie for our lasting friendship and connection across the oceans. Finally, I want to thank Karen and Morris Trabsky who have come to signify home to me.

Marc Trabsky has been the most attentive and critical reader since the early days of my academic career. He has patiently discussed each idea in detail, read every draft, and provided encouragement and support at every step. I am immensely thankful for his insight, care and companionship.

Parts of this book have been published elsewhere, and I am grateful for the permission to include those publications in this book: an early version of chapter five in "The Problem of Sexual Imbalance and Techniques of the Self in the *Diagnostic and Statistical Manual of Mental Disorders*," *History of Psychiatry* 27, no. 3 (2016): 320–335; and an early version of chapter six in "Pharmaceutical Intimacy: Managing Female Sexuality through Addyi," *Sexualities* 21, no. 4 (2018): 569–586.

Contents

CHAPTER 1

A Cartography of Appetites

This book offers a genealogy of the medicalisation of sexual appetite in Europe and the United States from the nineteenth to twenty-first centuries. Histories of sexuality have predominantly focused on the emergence of sexual identities and categories of desire. They have marginalised questions of excess and lack, the appearance of a libido that dwindles or intensifies, which became a pathological object in Europe by the nineteenth century. Through a genealogical approach that draws on the writings of Michel Foucault, *A Genealogy of Appetite in the Sexual Sciences* examines key "moments" in the pathologisation of sexuality and demonstrates how medical techniques assumed critical roles in shaping modern understandings of the problem of appetite. It examines how techniques of the patient case history, elixirs and devices, measurement, diagnostic manuals and pharmaceuticals were central to the medicalisation of sexual appetite. The book argues that these techniques are significant for understanding how a concern with "how much?" has transformed medical knowledge of sexuality since the nineteenth century. The questions of "how much?," "how often?" and "how intense?" thus require a genealogical investigation that pays attention to the emergence of medical techniques, the transformation of forms of knowledge and their effects on the problematisations of sexual appetite.

J. Flore, *A Genealogy of Appetite in the Sexual Sciences*,
https://doi.org/10.1007/978-3-030-39423-3_1

The Problem of Appetite

In the opening pages of *The Will to Knowledge* (1978), Michel Foucault remarks that speaking about sexuality as repressed has a seductive appeal. It enables one to believe they are subverting the social and political order with the inflammatory discourse of revolution. However, *The Will to Knowledge* was written in part to critique ideas of sexual liberation.[1] Foucault famously refutes the hypothesis that bourgeois societies had been governed by a regime of repression until the twentieth century. Instead, he suggests that there were two different procedures for expanding knowledge on sexuality in the West. The first, *scientia sexualis*, had at its centre the classification of types of pathologies and the professionalisation of a field of knowledge on sexuality. The second procedure, which Foucault attributes to "China, Japan, India, Rome, [and] the Arabo-Moslem societies," was *ars erotica*.[2] Erotic art was a creative, aesthetic and relational activity and a technique for the experiential transmission of knowledge of sensuality. Knowledge was communicated not through doctrine, but "in an esoteric manner and as the culmination of an initiation in which [the master] guides the disciple's progress with unfailing skill and severity."[3] While the erotic arts were certainly taught and learnt, there was no established institution governing the practice. It was an *embodied* experience transmitting the knowledge of a somatic relation through ritual and initiation.

Foucault conceives of *ars erotica* in terms of "unregulated" sensuality: "truth is drawn from pleasure itself, understood as a practice and accumulated as experience."[4] In other words, *ars erotica* is irreducible to "sexuality," and it cannot be constrained by the classificatory functions of *scientia sexualis*. In *The Will to Knowledge*, Foucault writes that the erotic arts did not conceptualise pleasure in medical, juridical or prohibitive terms. The identity of objects was less important than the experience of pleasure, an experience that could be enhanced and prolonged, augmented and stymied.

[1] Didier Eribon, "Michel Foucault's Histories of Sexuality," *GLQ: A Journal of Gay and Lesbian Studies* 7, no. 1 (2001): 43. See also Kevin Floyd, "Rethinking Reification: Marcuse, Psychoanalysis, and Gay Liberation," *Social Text* 19, no. 1 (2001): 103–126.

[2] Michel Foucault, *The Will to Knowledge: The History of Sexuality, Volume 1*, trans. Robert Hurley (London: Penguin, 1978), 57.

[3] Ibid.

[4] Ibid.

If pleasure needed to be controlled, the purpose was always to *intensify* the experience.

The *ars erotica/scientia sexualis* dichotomy has had a profound influence on constructivist histories of gender and sexuality. Contemporary scholars have been particularly interested in the genealogy of *scientia sexualis* in the West and its continued manifestations.[5] Criticism of Foucault's conceptual manoeuvre has centred on the purported distinction between art and science,[6] Foucault's "orientalism" of the East[7] and the question of whether Western societies ever possessed traditions akin to *ars erotica*.[8] Few works, however, have questioned the purported disappearance of *ars erotica* from Western discourses of sexuality. Furthermore, there has been scarce commentary on how Foucault further complicates the *ars erotica/scientia sexualis* disjunction in *The Will to Knowledge*:

> *ars erotica* did not disappear altogether from Western civilization; nor has it always been absent from the movement by which one sought to produce a science of sexuality… we must ask whether, since the nineteenth century, the *scientia sexualis*—under the guise of its decent positivism—has not functioned, at least to a certain extent, as an *ars erotica*. Perhaps this production of truth, intimidated though it was by the scientific model, multiplied, intensified, and even created its own intrinsic pleasures … We have at least invented a different kind of pleasure: pleasure in the truth of pleasure, the

[5] See, for example, Jan Bremmer, ed., *From Sappho to De Sade: Moments in the History of Sexuality* (London: Routledge, 1991), Chiara Beccalossi, *Female Sexual Inversion: Same-Sex Desires in Italian and British Sexology, c. 1870–1920* (New York: Palgrave Macmillan, 2012), and Patricia Caplan, ed., *The Cultural Construction of Sexuality* (New York: Routledge, 1987).

[6] For example, Mark Johnson, "Sexuality" in *Cultural Geography: A Critical Dictionary of Key Concepts*, eds. David Atkinson, Peter Jackson, David Sibley and Neil Washbourne (London: I.B. Tauris, 2005), 122–127, and Leon Antonio Rocha, "Scientia Sexualis Versus Ars Erotica: Foucault, van Gulik, Needham," *Studies in History and Philosophy of Biological and Biomedical Sciences* 42 (2011): 328–343.

[7] See the chapters by Valerie Traub and Dina Al-Kassim in *Islamicate Sexualities: Translations across Temporal Geographies of Desire*, eds. Kathryn Babayan and Afsaneh Najmabadi (Cambridge, MA: Harvard University Press, 2008), 1–40 and 297–340 and Gregory M. Pflugfelder, *Cartographies of Desire: Male-Male Sexuality in Japanese Discourse, 1600–1950* (Berkeley: University of California Press, 1999), and Jonathan Burton, "Western Encounters with Sex and Bodies in Non-European Cultures, 1500–1700" in *The Routledge History of Sex and the Body: 1500 to Present*, eds. Sarah Toulalan and Kate Fisher (London and New York: Routledge, 2013), 495–510.

[8] Romana Byrne, *Aesthetic Sexuality: A Literary History of Sadomasochism* (New York: Bloomsbury, 2013).

> pleasure of knowing that truth ... all this constitutes something like the errant fragments of an erotic art that is secretly transmitted by confession and the science of sex.[9]

Foucault identifies a form of *ars erotica* in the relentless production of truth central to *scientia sexualis.* He does not provide a more detailed explanation of how *ars erotica* might have continued to exist in Western discourses of sexuality; however, this excerpt sheds light on the importance of pleasure in Foucault's genealogy of the science of sexuality. Therein lies the critical relevance of *ars erotica* for a genealogy of sexual appetite. In this book, I suggest that *ars erotica*, in the form of the problematisation of appetite,[10] endures *within* the science of sexuality itself. Indeed, one of the main arguments of this book is that appetite has structured, to various degrees, the science of sexuality, from its inception to contemporary times.

In an interview conducted in 1977, Foucault acknowledges that the erotic arts were, to a certain extent, medicalised in so far as they utilised "the means (pharmaceutical or somatic) which serve to intensify pleasure."[11] Medicalisation in *ars erotica* focused on methods for maximising pleasures, while in *scientia sexualis*, it deployed an apparatus of sexuality where identity became articulated with reference to objects of desire. The chapters that follow will problematise this disjuncture by demonstrating how the management of pleasure is deeply enmeshed in the science of sexuality. As histories of sexualities have demonstrated since the 1970s, object choice became the privileged focus of sexual science in the West. The medicalisation of sexuality produced a "precise categorization of a sexual behavior linked to psychology, linked to a desire."[12]

[9] Foucault, *The Will to Knowledge*, 70–71.

[10] Problematisation is "the set of discursive and nondiscursive practices that makes something enter into the play of the true and false, and constitutes it as an object for thought." Michel Foucault, "The Concern for Truth," in *Foucault Live (Interviews, 1961–1984)*, ed. Sylvère Lotringer, trans. Lysa Hochroth and John Johnston (New York: Semiotext(e), 1996), 456–457. See also Michel Foucault, "Polemics, Politics, and Problematizations: An Interview with Michel Foucault" in *Ethics: Essential Works of Foucault 1954–1984, Volume 1*, ed. Paul Rabinow (London: Penguin Books, 1997), 114.

[11] Michel Foucault, *Power/Knowledge: Selected Interviews and Other Writings, 1972–1977* (New York: Vintage Books, 1980 [1976]), 191.

[12] Michel Foucault, "The Gay Science," *Critical Inquiry* 37 (2011 [1978]): 387. In an interview in 1984, Foucault also notes, "for centuries people generally, as well as doctors, psychiatrists, and even liberation movements, have always spoken about desire, and never

In *The Use of Pleasure* (1984), Foucault argues that pleasure emerged as a problem of ethical conduct for the Ancient Greeks and Romans. It was "a matter of regimen aimed at regulating an activity that was recognized as being important for health."[13] The question of pleasure turned its use into an ethical task and an exercise in moderation. Prudence was advised against an excessive enjoyment of passion and sensuality. The individual had to cultivate their mind and body in order to achieve an ethical use of pleasure. The second volume of *The History of Sexuality* thus expands on Foucault's earlier writings on ethics by offering a genealogy of how individuals came to stylise themselves as ethical subjects of sexual behaviour.[14] Foucault shifts his attention though to Greek and Roman antiquity in order to determine how "Western man had been brought to recognize himself as a subject of desire."[15] The volume does not position *ars erotica* as identical to the Ancient Greek's *chrēsis aphrodisiōn* (use of pleasures).[16] However, it reveals how the amount and dynamic of pleasure were historically problematised, and how such problematisations became integral to the formation of ethical conduct.

The problematisation of *aphrodisia* in Greek and Roman antiquity focused on an economy of amounts, rather than an economy of deviancy and pathology. Individuals were counselled to use pleasure "more amply ... or in smaller amounts ... or as little as possible."[17] Foucault explains that the individual needed to take care when partaking in amorous activities because of the different sensations produced by their body. Those

about pleasure. 'We have to liberate our desire,' they say. *No!* We have to create new pleasure. And then maybe desire will follow." Michel Foucault "Sex, Power, and the Politics of Identity" in *Ethics: Essential Works of Foucault 1954–1984, Volume 1*, ed. Paul Rabinow (London: Penguin Books, 1997), 166 (emphasis original).

[13] Michel Foucault, *The Use of Pleasure: The History of Sexuality, Volume 2*, trans. Robert Hurley (New York: Vintage, 1985 [1984]), 97–98.

[14] Ibid., 32. See further Timothy O'Leary, *Foucault and the Art of Ethics* (London and New York: Continuum, 2002), 43.

[15] Ibid., 6.

[16] Foucault, *The Use of Pleasure*, 32. Foucault suggests that *aphrodisia* referred to complex dynamics of acts, pleasure and desire. While irreducible to sexuality, the problematisation of *aphrodisia* was conceptualised in terms of quantity and occasion. It was connected to its uses (*chrēsis*), and this conditioned its emergence as a problem, where seasons, time and situation became crucial factors in advice on when to engage in intimate relations. Acts were not forbidden because they were deemed abnormal; they were problematised because of their intensity, context and quantity.

[17] Foucault, *The Use of Pleasure*, 114–115.

sensations "raised or lowered the level of each of the elements that were responsible for the body's equilibrium"[18]—a reference to the humoral model of the body.[19] The problem of too much or too little, according to Foucault, was managed in Greek and Roman antiquity through the development of techniques of dietetics. The word "diet" encompassed various areas of life in addition to nourishment and carnal relations, such as exercise and sleep. As such, dietetics was understood broadly as a set of techniques for conceptualising the dynamics of the body, its needs and its relations. It was part of an ensemble of philosophical and medical tools for achieving self-control in the use of pleasures.

This book builds on Foucault's genealogy of *aphrodisia* in *The Use of Pleasure* by asking *how* appetite became problematised in Europe and the United States in the nineteenth to twenty-first centuries. It examines how the medicalisation of sexual appetite emerged at different historical moments in modern medico-scientific discourses on sexuality and contends that the intensification or dwindling of appetite was never separate from but rather *integral* to the science of sex. In other words, I argue that a history of the use of pleasures is *not* mutually exclusive from a history of *scientia sexualis. A Genealogy of Appetite in the Sexual Sciences* approaches the development of knowledge on sexual appetite and their fluctuations as mediated by medical techniques. It examines techniques that transform knowledge and represent moderation as the most desirable and ideal form of the sexual self. Indeed, it is only by taking seriously the role assumed by techniques of the patient case history, elixirs and devices, measurement, diagnostic manuals and pharmaceuticals in the medicalisation of sexual appetite that we can understand how the discourse of the use of pleasure continues to affect the evolution of medico-scientific ideas on the management of sexuality.

By rethinking the history of sexuality *through* a history of sexual appetite, this book is situated within a genealogical approach to the study of ideas. This methodology draws attention to the "series of interpretations

[18] Ibid., 115. Lesley Dean-Jones also writes that *gnothi seauton* ("know yourself") and *meden agan* ("nothing to excess") were considered central to Ancient Greek thought. See Lesley Dean-Jones, "The Politics of Pleasure: Female Sexual Appetite in the Hippocratic Corpus" in *Discourses of Sexuality: From Aristotle to AIDS*, ed. Domna C. Stanton (Ann Arbor: University of Michigan Press, 1992), 50.

[19] See Dean-Jones, "The Politics of Pleasure."

[which] have their own historical specificity."[20] While Foucault employed a genealogical method to examine the circulation of discourses on sexuality in *The Will to Knowledge*, I harness it to examine the changes, continuities and discontinuities, of the emergence of ideas on sexual appetite and the medical techniques that correspond to their problematisation. Far from seeking a "truth" to the problem of sexuality, the book argues that appetite cannot be considered as mutually exclusive from object choice, nor can it be reduced to a specific disorder or syndrome. The purpose of this genealogy then is not to locate an origin of sexual appetite or privilege its emergence in a history of sexuality, but rather to show how problems of sex and appetite inform each other and are conditioned by broader concerns with rhythms and fluctuations of desire, pleasure and intimacy.

Writing a Genealogy of Sexual Appetite

In *What is Sexual History?* Jeffrey Weeks argues that the modern West "has been preoccupied with *whom* people had sex with," in contrast to other periods which were concerned with "questions of excess or overindulgence, activity or passivity, sin or salvation."[21] Twentieth-century scholarship in sexuality studies has focused almost exclusively on questions of sexual rights, identity and norms, topics which typically have sexual object choice at their centre.[22] Sociological studies on homosexuality surfaced in the late 1960s, an approach that owed much to Alfred Charles Kinsey and Katharine Bement Davis (discussed in Chap. 4). In 1967, William Simon and John H. Gagnon drew on labelling theory and argued that the homosexual "has all of his acts interpreted through the framework of his homosexuality."[23] They later used the concept of "sexual scripts" to theorise how individuals learn and apply socio-sexual norms and

[20] Jennifer Germon, *Gender: A Genealogy of an Idea* (New York: Palgrave Macmillan, 2009), 14.

[21] Jeffrey Weeks, *What is Sexual History?* (Cambridge: Polity, 2016), 78 (emphasis original).

[22] See chapter one in Stephen Garton, *Histories of Sexuality: Antiquity to Sexual Revolution* (London: Routledge, 2004).

[23] William Simon and John H. Gagnon, "Homosexuality: The Formulation of a Sociological Perspective," *Journal of Health and Social Behavior* 8, no. 3 (1967): 177.

behaviour.[24] The concept of scripting drew from ideas of symbolic interactionism and constructionism—a challenge to Freudian notions of "instinct" and "the unconscious."[25] Feminist writer Mary McIntosh added to this scholarship by arguing that the labelling of homosexuals operates as a "mechanism of social control" facilitating the social segregation of so-called deviants, thus enabling "the bulk of society [to remain] pure."[26] McIntosh was already anticipating Foucault by arguing that homosexual identities began to emerge in England at the end of the seventeenth century.[27]

In activist and scholarly circles in the 1970s, prominent figures such as Weeks, John D'Emilio and Dennis Altman wrote histories of sexuality amid calls for liberation and focused their writings on the subject of gay oppression.[28] This also included the emergence and expansion of subcultures and communities around a contentious politics of sexual identity, intersected by race, class and gender. The formation of identity and demands for equal rights were central to the histories of sexuality written in the 1970s and 1980s. Scholars and activists were concerned with documenting silenced histories of gay men, lesbians, bisexuals and transgender people while also demanding the decriminalisation of homosexuality and its removal from the *Diagnostic and Statistical Manual of Mental Disorders*.[29] Thus, drawing on symbolic interactionism, labelling theory

[24] John H. Gagnon and William Simon, *Sexual Conduct: The Social Sources of Human Sexuality* (New Brunswick: Aldine Transaction, 2005 [1973]).

[25] See Sigmund Freud, *The Standard Edition of the Complete Psychological Works of Sigmund Freud, Volume VII*, trans. James Strachey (London: Vintage Books, 2001 [1905]).

[26] Mary McIntosh, "The Homosexual Role," *Social Problems* 16, no. 2 (1968): 184.

[27] It should be noted that McIntosh, in contrast to Foucault, does not locate the emergence of homosexuality in medical discourse.

[28] See, for example, Jeffrey Weeks, *Coming Out: Homosexual Politics from the Nineteenth Century to the Present* (London: Quartet Books, 1977), Dennis Altman, *Homosexual: Oppression and Liberation* (New York: New York University Press, 1993 [1971]), John D'Emilio, *Sexual Politics, Sexual Communities: The Making of a Homosexual Minority in the United States, 1940–1970* (Chicago: University of Chicago Press, 1983), and Vern L. Bullough, ed., *Before Stonewall: Activists for Gay and Lesbian Rights in Historical Context* (New York: Routledge, 2008). Marc Stein notes that transgender rights were also incorporated into the movement, although some groups did not consider themselves connected to the gay and lesbian rights coalition. Marc Stein, *Rethinking the Gay and Lesbian Movement* (Cambridge: Routledge, 2012), 152.

[29] See Ronald Bayer, *Homosexuality and American Psychiatry: The Politics of Diagnosis* (Princeton: Princeton University Press, 1987 [1981]) and Jack Drescher and Joseph

and social constructionism, the works of academics and activists challenged essentialist and biological understandings of sexual identity.[30]

Following the publication of the English translation of *The Will to Knowledge* in 1978, studies of sexuality became increasingly informed by Foucault's concept of "effective history," a form of historicism "without constants" and one which "deprives the self of the reassuring stability of life and nature," and is inextricable from questions of power, discourse and institutions.[31] Foucault terms this methodology *genealogy*. A genealogical approach to history focuses on "institutional and resistant operations of power within systems of thought."[32] Closely aligned to his concept of problematisation, genealogy enquires into how practices and behaviours become "problems" in society. For Foucault, knowledge intersects with relations of power upon which it depends to attain the status of "truth."[33] As a result, an idea or behaviour emerges as a norm due to continual historical struggles over meaning and knowledge. A genealogical method is sceptical of grand narratives, searches for truths or origins, and teleological understandings of progress and enlightenment.[34] Its focus is on *conditions of emergence*. Genealogy, then, is a form of "effective history" where social and economic conditions are as important as the formal knowledge produced by institutions. In developing and applying a genealogical methodology, Foucault aspires to write a "history of the present"[35]—an interrogation of present conditions through critically studying the past.

The scholarly attention to the formation of sexual identities in the nineteenth and twentieth centuries greatly influenced genealogies of

P. Merlino, eds., *American Psychiatry and Homosexuality: An Oral History* (New York: Harrington Park Press, 2007).

[30] See by Ken Plummer, "Awareness of Homosexuality," in *Contemporary Social Problems in Britain*, eds. Roy Bailey and Jock Young (Hants and Massachusetts: Saxon Books and Lexington Books, 1973), 103–125 and *Sexual Stigma: An Interactionist Account* (London: Routledge & Kegan Paul, 1975).

[31] Michel Foucault, "Nietzsche, Genealogy, History" in *Language, Counter-Memory, Practice: Selected Essays and Interviews by Michel Foucault*, ed. Donald F. Bouchard (New York: Cornell University Press, 1977), 153–154.

[32] Lisa Downing, *The Cambridge Introduction to Michel Foucault* (Cambridge: Cambridge University Press, 2008), 14.

[33] Foucault, *Power/Knowledge*, 93.

[34] Downing, *The Cambridge Introduction to Michel Foucault*, 15.

[35] See Michel Foucault, *Discipline and Punish: The Birth of the Prison*, trans. Alan Sheridan (Middlesex: Penguin, 1991 [1977]), 31. On the history of the present, see Jan Goldstein, ed., *Foucault and the Writing of History* (Oxford: Blackwell, 1994).

sexuality in the 1980s and 1990s, which turned from a history of struggle, rights and liberation to the discursive foundations of "types" of individuals. Historians offered new accounts of sexuality based on social constructionist understandings of power, knowledge and language, which were dominated by identity politics and organised by unravelling the hetero/homosexual binary.[36] Debates over essentialism and social constructionism,[37] or whether homosexuality was inherently biological or socially produced, tended to revolve around the politics of acquiring more rights for sexual minorities, such as freedom from prosecution, access to health care and rights for people living with HIV, while recognising fundamental differences between homosexuals and heterosexuals. Those fundamental differences however were not without contestation. David Halperin, for example, argues that "homosexuality itself" and "heterosexuality itself" are flawed categories as they do not constitute "independent modes of sexual being, leading some sort of ideal existence apart from particular human societies, outside of history and culture."[38] While the work of Halperin and others consider the complexity of questions of acts and identities in a genealogy of sexuality, their conceptualisations has generally overlooked the dynamics of those acts, the contingencies of appetite, and the problematisations of amounts in the science of sex.[39]

In the mid-to-late 1990s, another influential paradigm for writing a genealogy of sexuality emerged in the form of queer theory.[40] With roots

[36] See Luce Irigaray, *Ce sexe qui n'en est pas un* (Paris: Éditions de Minuit, 1977), Adrienne Rich, "Compulsory Heterosexuality and Lesbian Existence," in *The Lesbian and Gay Studies Reader*, eds. Henry Abelove, Michèle Aina Barale and David M. Halperin (New York: Routledge, 1993 [1980]), 227–254, Sylvère Lotringer, *Overexposed: Perverting Perversions* (New York: Semiotext(e), 1988), François Peraldi, ed., *Polysexuality* (New York: Semiotext(e), 1981), and Monique Wittig, *The Straight Mind and Other Essays* (Boston: Beacon Press, 1992).

[37] See Steven Epstein, "Gay Politics, Ethnic Identity: The Limits of Social Constructionism," *Socialist Review* 93/94 (1987): 9–54.

[38] David Halperin, *One Hundred Years of Homosexuality and Other Essays on Greek Love* (New York: Routledge, 1990), 45.

[39] See, for example, Jonathan Ned Katz, *The Invention of Heterosexuality* (New York: Dutton, 1995).

[40] See Eve Kosofsky Sedgwick, *Epistemology of the Closet* (Berkeley: University of California Press, 1990), Judith Butler, *Gender Trouble: Feminism and the Subversion of Identity* (New York: Routledge, 1990), Teresa De Lauretis, *Queer Theory: Lesbian and Gay Sexualities* (Bloomington: Indiana University Press, 1991), Michael Warner, ed., *Fear of a Queer Planet: Queer Politics and Social Theory* (Minneapolis: University of Minnesota Press, 1993), Brett Beemyn and Mickey Eliason, eds., *Queer Studies: a Lesbian, Gay, Bisexual and Transgender*

in activist movements, the term "queer" remains contested in academic and activist circles.[41] Nonetheless, drawing on post-structuralism, queer theory subjected "regimes of the normal"[42] to critical enquiry, opting for "denaturalisation as its primary strategy."[43] Combining an ethos of resistance and protest, as well as a commitment to critique authoritative knowledge, whether scientific, literary or historical, queer theory destabilises notions of identity and approaches sexuality as fluid, changing, and importantly, performative.[44] The historiographies of sexuality that draw on the insights of queer theory approach historical documents as cultural, political and socio-economic artefacts, "fusing the work of excavation with the recognition that sexualities are socially constructed and can take multiple forms."[45] These important works have reframed how histories of queer sexualities are narrated and revealed the instability of identities as well as their epistemological contingency.[46] Yet, if queer theory aims to denaturalise categories of knowledge, scholars have had little to say about its potential for problematising dimensions of appetite in sexuality, not only in terms of their history, but also to "queer" flows of appetite itself.[47]

Anthology (New York: New York University Press, 1996), Steven Seidman, ed., *Queer Theory/Sociology* (Malden, MA: Blackwell, 1996), Elizabeth Weed and Naomi Schor, eds., *Feminism Meets Queer Theory* (Bloomington: Indiana University Press, 1997), Fabio Cleto, ed., *Camp: Queer Aesthetics and the Performing of the Subject, A Reader* (Edinburgh: Edinburgh University Press, 1999), and José Esteban Muñoz, *Disidentifications: Queers of Color and the Performance of Politics* (Minneapolis: University of Minnesota Press, 1999).

[41] See Douglas Crimp, "Right On, Girlfriend!" in *Fear of a Queer Planet: Queer Politics and Social Theory*, ed. Michael Warner (Minneapolis: University of Minnesota Press, 1993), 300–320 and Steven Maynard, "'Respect Your Elders, Know Your Past': History and the Queer Theorists," *Radical History Review* 75 (1999): 56–78.

[42] Michael Warner, "Introduction," in *Fear of a Queer Planet: Queer Politics and Social Theory*, ed. Michael Warner (Minneapolis: University of Minnesota Press, 1993), xxvi.

[43] Annamarie Jagose, *Queer Theory* (Melbourne: Melbourne University Press, 1996), 98.

[44] See Butler, *Gender Trouble.*

[45] Susan McCabe, "To Be and to Have: The Rise of Queer Historicism" (Book review), *GLQ: A Journal of Lesbian & Gay Studies* 11, no. 1 (2005): 121. See also Valerie Traub, "The New Unhistoricism in Queer Studies," *PMLA* 128, no. 1 (2013): 21–37.

[46] See William B. Turner, *A Genealogy of Queer Theory* (Philadelphia: Temple University Press, 2000), Carla Freccero, *Queer/Early/Modern* (Durham: Duke University Press, 2006), and Lynne Huffer, *Mad for Foucault: Rethinking the Foundations of Queer Theory* (New York: Columbia University Press, 2010).

[47] While a number of genealogies of sexuality have shown how identity formation emerges in the nineteenth and twentieth centuries amid the pathologisation of sexuality as both excessive and lacking, in relying on (often binary) paradigms of identity, they have neglected other axes of analysis such as the problematisation of appetite in itself. See, for example, Steven

In recent years, there has been a resurgence of histories of sexuality focused on tracing the genealogies of disorders of appetite. In their intellectual history of frigidity, Peter Cryle and Alison Moore examine the multiple, conflicting meanings of frigidity from the seventeenth to twentieth century. They propose the term "sexual coldness" to encompass the varying definitions of the medical and societal category of frigidity. The discursive trajectories of this term reveal its importance in a medical history of sexuality. Frigidity was at times considered a precursor to sexual inversion (homosexuality) and often defined as an inability to reproduce or to engage in heterosexual relations. It slipped between a physical and a psychological condition and was connected to many other "disorders" such as hysteria, nymphomania and neurasthenia.[48] On the other hand, Angus McLaren's cultural history of impotence in the Western world unpacks the historical connections between a lack of desire in men, virility, infertility and changing models of masculinity, with a focus on the shift in the meanings of impotence from infertility to erectile dysfunction. As do Cryle and Moore, McLaren draws attention to the connection between discourses of indulgence and restraint in matters of sexuality.[49] Further, McLaren observes that in the nineteenth and early-twentieth centuries, the problem of low desire in men (often associated with neurasthenia) gave rise to quacks and various kinds of elixirs for the reviving of desire[50]—a phenomenon that I discuss in Chap. 3 of this book.

On the other end of the spectrum of sexual appetite, Carol Groneman observes that disorders of excess were understood to affect men and women (for men, satyriasis), though women's "irrationality" was key to

Angelides, *A History of Bisexuality* (Chicago: University of Chicago Press, 2001), Kathryn R. Kent, *Making Girls into Women: American Women's Writing and the Rise of Lesbian Identity* (Durham: Duke University Press, 2003) and Chiara Beccalossi, *Female Sexual Inversion: Same-Sex Desires in Italian and British Sexology, c. 1870–1920* (New York: Palgrave Macmillan, 2012).

[48] Peter Cryle and Alison Moore, *Frigidity: An Intellectual History* (London: Palgrave Macmillan, 2011). Much has been written on the history of hysteria, see, for example, Cristina Mazzoni, *Saint Hysteria: Neurosis, Mysticism, and Gender in European Culture* (New York: Cornell University Press, 1996) and Georges Didi-Huberman, *Invention of Hysteria: Charcot and the Photographic Iconography of the Salpêtrière*, trans. Alisa Hartz (Cambridge, MA: MIT Press, 2003).

[49] Angus McLaren, *Impotence: A Cultural History* (Chicago and London: University of Chicago Press, 2007).

[50] Ibid., 126–148.

medical discourses on nymphomania at the time.[51] The intensity of women's desires was considered a threat to a society of which they were the moral guardians: "men have more sexual desire, but less disease of excess; women are less desirous, but more prone to morbid passions."[52] A key historical text of relevance here is M. D. T De Bienville's *La nymphomanie* (1771) which argued that the problem lay in the *intensity* of women desiring sexual satisfaction, their genitals demanding attention and their minds overwhelmed by lascivious images.[53] Groneman's important intervention highlights that by the twentieth century, "the theory that nymphomaniacs were actually frigid and did not experience orgasm, thus their 'insatiability'"[54] gained increasing currency. This, for Groneman, culminates in the development of the category of sexual dysfunctions in psychiatry in the United States, which I will examine in Chaps. 4 and 5 of this book. Conversely, Barry Reay, Nina Attwood and Claire Gooder trace the manifestations of excess as "sex addiction" in popular culture and medical treatises in the twentieth and twenty-first centuries. Their historical intervention contends that sexual hyperaesthesia in the nineteenth century affected more men than women, while sexual anaesthesia was considered a woman's problem. Reay, Attwood and Gooder assert that while nineteenth-century sexology "has documented and discussed the concept of excessive sexual desire and behaviour," this was done "to a very limited extent."[55] Yet as I demonstrate in Chap. 2, the problematisation of excess

[51] Carol Groneman, "Nymphomania: The Historical Construction of Female Sexuality," *Signs* 19, no. 2 (1994): 345. On the history of satyriasis, see Timothy Verhoeven, "Pathologizing Male Desire: Satyriasis, Masculinity, and Modern Civilization at the Fin de Siècle," *Journal of the History of Sexuality* 24, no. 1 (2015): 25–45.

[52] M D T de Bienville, *La nymphomanie, ou traité de la fureur utérine* (Paris: Office de Librairie, 1886 [1771]), 352.

[53] Replicating the humoral model of the human body into the eighteenth century, the disease of heat for women was known as "uterine fury," evoking the idea of a combustible, unruly uterus. Excessive appetite served as an organising element of sexual aberrations and was considered the "ultimate" form of depravity. In *La folie érotique* (1888), for example, Benjamin Ball attempted to classify ailments of sexual excess in men and women. The table lists "Erotic Madness" as the overarching category, with "erotomania," "sexual excitement" and "sexual perversions" as subcategories. Under "sexual excitation," he listed nymphomania and satyriasis, and under "sexual perversion," he included necrophiliacs, pederasts and inverts. Benjamin Ball, *La folie érotique* (Paris: J B Ballière, 1888), 9 (translation author).

[54] Groneman, "Nymphomania," 359.

[55] Barry Reay, Nina Attwood and Claire Gooder, *Sex Addiction: A Critical History* (Cambridge: Polity, 2015), 20.

was far from limited in sexological treatises in the nineteenth century. Indeed, this book will show how appetite, whether excessive or lacking, was conspicuous within the pathologisation of sexual object choice, and at times deemed more problematic, in a range of psychiatric and medical treatises from the nineteenth to twenty-first centuries.

A Genealogy of Appetite in the Sexual Sciences

In writing a genealogy of the medicalisation of sexual appetite from the nineteenth to twenty-first centuries, this book focuses on the development of Western psychiatry in Germany, France, England and the United States. It examines the role that psychiatric institutions assumed during this period in producing and disseminating discourses of sexuality and the medical techniques that corresponded to the problematisation of sexual appetite. The term "technique" derives from the Greek *technē* meaning, in different contexts, "art" or "craft." Foucault's later works focus on the centrality of techniques for governance and the formation of subjectivity. Foucault refers to *technē* as *savoir-faire*, meaning "know-how" or practice.[56] His exploration of technique is influenced by Marcel Mauss' identification of the body as one's "first and most natural instrument."[57] For Mauss, the body is a technical object, such that human modes of action in the social world are always already mediated by techniques. To put this differently, techniques have a fundamentally social dimension; they mediate how bodies operate in everyday life, which come to reflect norms and cultural mores.

This book's interest in medical techniques created, deployed and harnessed by psychiatric institutions and their role in shaping modern understandings of the problem of sexual appetite is also influenced by Nikolas Rose's study of the "*techne* of psychology" that is, "the ways in which it is organized as a practice."[58] Rose's study of psychology is relevant here because he approaches the field as being deployed around "*problems*, exercising a certain *diagnostic* gaze, grounded in a claim to *truth*, asserting technical *efficacy*, and avowing *humane* ethical virtues."[59] Psychiatry, like

[56] Foucault, *The Use of Pleasure*, 62.

[57] Marcel Mauss, "Techniques of the Body," *Economy & Society* 2, no. 1 (1973): 77.

[58] Nikolas Rose, *Inventing Our Selves: Psychology, Power and Personhood* (Cambridge: Cambridge University Press, 1998), 85.

[59] Ibid., 86 (emphases original).

psychology, harnesses a range of techniques for the diagnosis, classification and treatment of "patients." The technical deployment of knowledge is not characterised by the oppression of the subject, but rather conditions the emergence of the subjectivity of the patient. Within this "apparatus,"[60] the patient is always connected to institutional relations and implicated in a "game of power" (*jeu de pouvoir*).[61] They are embroiled in "a set of strategies of the relations of forces supporting, and supported by, certain types of knowledge."[62] The medical techniques that emerged in and around psychiatric institutions during the nineteenth to twenty-first centuries must be considered not only for conceptualising how sexual appetite was problematised but for how knowledge of appetite is continually transformed. The techniques of the patient case history, elixirs and devices, measurement, diagnostic manuals and pharmaceuticals reveal how psychiatry deploys power-knowledge relations with patients and disseminates discourses on sexual excess and lack, its pathologisation and the necessity of moderation in the enjoyment of sexual pleasures.

This book argues that a genealogy of the medicalisation of sexual appetite can only be understood by examining the role assumed by psychiatric institutions in managing the problem of appetite and the techniques that congealed around them. The medical techniques examined in the following pages are not incidental. They play a central role in the development of understandings of sexual appetite in psychiatric institutions from the nineteenth to twenty-first centuries. Chapter 2 opens with an examination of sexual appetite in nineteenth-century sexology and the technique of the patient case history. Sexology in Germany, France and England during this period is recognised as a significant moment in the emergence of psychiatric knowledge of sexuality in the West.[63] It developed as a field of knowledge chiefly concerned with the pathologisation of sexual object choice. This chapter examines how the medicalisation of sexual appetite in nineteenth-century sexology emerged *through* the technique of the patient

[60] "Apparatus" is the English translation of the term *dispositif* used in Foucault's works. In *The Will to Knowledge*, he connects apparatus to the deployment of sexuality (106).

[61] Michel Foucault, "Le jeu de Michel Foucault," in *Michel Foucault: Dits et écrits, 1954–1988, tome III 1976–1979* (Paris: Gallimard, 1994 [1977]), 300.

[62] Foucault, *Power/Knowledge*, 94–96.

[63] See Gert Hekma, "A History of Sexology: Social and Historical Aspects of Sexuality," in *From Sappho to De Sade: Moments in the History of Sexuality*, ed. Jan N. Bremmer (London: Routledge, 1991), 173–193 and Arnold I. Davidson, *The Emergence of Sexuality: Historical Epistemology and the Formation of Concepts* (Cambridge, MA: Harvard University Press, 2001).

case history. It considers two aspects of this technique: first, how the case history presented sexual appetite as a structuring device in the expansion of taxonomies of sexual perversions, and second, how this was accomplished by inextricably tying the imagination and narrative to the notion of sexual excess. The imagination formed the bedrock of sexuality itself and was treated as both essential and suspicious. The patient case history was a discursive device linking pathology, excess and the imagination. It constituted a technique for the ordering of knowledge on sexual appetite and its dissemination.

The genealogy of sexual appetite however cannot solely be written through "formal" annals such as nineteenth-century sexological treatises. Previous genealogies of sexuality have drawn on wider archives including advice literature from experts and non-experts and fictional texts.[64] Turning to the late-nineteenth and early-twentieth centuries in the United States—a particularly fertile era for the production of techniques and discourses of sexual appetite—Chap. 3 opens a different archive. It traces the circulation of elixirs and devices marketed by quacks and moral educators to "heal" so-called lost manhood. It examines the commodification of sexual appetite through the advertisements of elixirs and mechanical devices in manuals, pamphlets, tracts and newspapers. The subject who emerged at the beginning of the twentieth century was not simply a patient, but a customer who needed both a product and knowledge to manage and control their sexuality. Patients were no longer considered passive subjects of medical diagnoses and instead became active consumers utilising a range of techniques to both enhance and govern their sexual appetites. To this extent, the chapter considers how the marketing of elixirs and mechanical devices for sexual imbalance were used for managing appetite. The different products promoted in this era functioned as techniques for actively encouraging individuals to autonomously and independently manage their sexual lives. Whether individuals were encouraged to consume foods, elixirs or devices, the narrative of responsibility and self-improvement permeated the marketing and use of those products.

The twentieth century in the United States, particularly following the movement of sexologists and psychiatrists from Europe to America after

[64] See, for example, Roy Porter "The Literature of Sexual Advice before 1800," in *Sexual Knowledge, Sexual Science*, eds. Roy Porter and Mikuláš Teich (Cambridge: Cambridge University Press, 1994), 134–157, and Roy Porter, *A Social History of Madness* (London: Weidenfeld and Nicolson, 1987).

World War II, witnessed the emergence of the modern psychiatric institution.[65] Foundational to the development of sexual science in the twentieth century were the works of Alfred Charles Kinsey, and William H. Masters and Virginia E. Johnson. Chapter 4 analyses how Kinsey utilised statistics and the concept of averages in his research on human sexuality. It argues that sexual appetite conditioned how statistical data was used in the Kinsey studies. The Kinsey team mobilised questions of "how much?" and "how often?" to produce graphs on which sexual appetite could be counted and mapped. Turning to the work of Masters and Johnson and the use of techniques of observation and measurement in the creation of norms of sexual behaviour, the chapter explores how the researchers further opened sexual activity to scientific investigation. Their work cemented norms of sexual appetite, presenting both the necessity of perfecting techniques to achieve pleasure and the norm to which individuals should aspire. This chapter thus contends that the works of Kinsey, and Masters and Johnson were important for reifying concepts of averages and norms and for developing techniques for the measurement of sexual appetite.

Building on the work of Kinsey and Masters and Johnson, Western psychiatry developed in the late-twentieth and early-twenty-first centuries an enhanced physiological model of human sexual response. Chapter 5 examines this discursive expansion in the classifications and interpretations of "sexual dysfunctions" in the United States. It explores how the turn of the psychiatric gaze towards amount, balance and frequency in the *Diagnostic and Statistical Manual of Mental Disorders* harnesses a range of techniques that, in addition to pathologising the patient, invite her to develop greater awareness of her sexual self. The patient is actively encouraged to use the manual as well as information gathered during the therapeutic process as

[65] The historian Gert Hekma explains that after the Second World War, "the United States became the centre of sexology, and with the research of the biologist Alfred Kinsey, it acquired a sociological character." Hekma, "A History of Sexology," 186. From Chap. 4, this book will only be focusing on the United States due to the breadth and depth of sexological research conducted by the team led by Kinsey, and the groundbreaking work of Masters and Johnson, which were foundational to twentieth-century psychiatric understandings of sexual appetite. However, it is important to note the research conducted in other countries that left a lasting imprint on psychiatry and sexology. For example, see Germon, *Gender: A Genealogy of an Idea* on the work of New Zealand-born John Money.

techniques for managing her sexual imbalances. This chapter thus aims to historicise the turn of the psychiatric gaze towards the problematisation of sexual appetite in the late-twentieth and early-twenty-first centuries, and, in doing so, to approach the development of the diagnostic manual as a technical object for both professional diagnosis and for the care of the self.

However, alongside revisions of the diagnostic manual, the late-twentieth and early-twenty-first centuries witnessed the intensification in psychiatry of the development and prescription of pharmaceuticals for sexual imbalance. Chapter 6 explores the emergence of Addyi (flibanserin) as a case study of how this technique produces a particular subject of pharmaceutical knowledge. The chapter considers the significance of the act of pharmaceutical ingestion on the embodied subjectivity of the consumer and the chemical constitution of the human body. The use of Addyi to manage sexual imbalance in combination with the diagnostic manual converges in the emergence of a socio-technical and knowledge-gathering subject. This subject is armed with tools to monitor the self and gather knowledge of her sexual appetite, a process that affirms intimacy. Operating alongside the diagnostic manual, Addyi purports to act on the chemical composition of the brain and embeds techniques of management of sexual appetite within the body itself. Chapter 6 thus explores how pharmaceutical intimacy as a technique of self-management represents sexual subjectivity mediated by written text, spoken word and chemical interactions.

In writing a genealogy of the medicalisation of sexual appetite in Europe and the United States from the nineteenth to twenty-first centuries, this book invites us to reflect on how amount, balance and frequency continue to be governed through psychiatric institutions. It asks us what possibilities can be excavated and generated if we think of a history of sexuality differently; if we understand desire and pleasure as not solely a matter of "whom?" but also, "how much?", "how often?" and "how intense?" To write a genealogy of sexuality through the lens of sexual appetite does not only involve interrogating how concepts of amount, balance and frequency were problematised in the past, but also examining how psychiatric institutions continue to question the use of pleasure and develop techniques for its medicalisation today.

Bibliography

Al-Kassin, Dina. 2008. Epilogue: Sexual Epistemologies, East in West. In *Islamicate Sexualities: Translations Across Temporal Geographies of Desire*, eds. Kathryn Babayan and Afsaneh Najmabadi, 297–340. Cambridge, MA: Harvard University Press.

Altman, Dennis. 1993 [1971]. *Homosexual: Oppression and Liberation*. New York: New York University Press.

Angelides, Steven. 2001. *A History of Bisexuality*. Chicago: University of Chicago Press.

Ball, Benjamin. 1888. *La folie érotique*. Paris: J B Ballière.

Bayer, Ronald. 1987 [1981]. *Homosexuality and American Psychiatry: The Politics of Diagnosis*. Princeton: Princeton University Press.

Beccalossi, Chiara. 2012. *Female Sexual Inversion: Same-Sex Desires in Italian and British Sexology, c. 1870–1920*. New York: Palgrave Macmillan.

Beemyn, Brett, and Mickey Eliason, eds. 1996. *Queer Studies: A Lesbian, Gay, Bisexual and Transgender Anthology*. New York: New York University Press.

Bremmer, Jan, ed. 1991. *From Sappho to De Sade: Moments in the History of Sexuality*. London: Routledge.

Bullough, Vern L., ed. 2008. *Before Stonewall: Activists for Gay and Lesbian Rights in Historical Context*. New York: Routledge.

Burton, Jonathan. 2013. Western Encounters with Sex and Bodies in Non-European Cultures, 1500–1700. In *The Routledge History of Sex and the Body: 1500 to Present*, eds. Sarah Toulalan and Kate Fisher, 495–510. London and New York: Routledge.

Butler, Judith. 1990. *Gender Trouble: Feminism and the Subversion of Identity*. New York: Routledge.

Byrne, Romana. 2013. *Aesthetic Sexuality: A Literary History of Sadomasochism*. New York: Bloomsbury.

Caplan, Patricia, ed. 1987. *The Cultural Construction of Sexuality*. New York: Routledge.

Cleto, Fabio, ed. 1999. *Camp: Queer Aesthetics and the Performing of the Subject, A Reader*. Edinburgh: Edinburgh University Press.

Crimp, Douglas. 1993. Right On, Girlfriend! In *Fear of a Queer Planet: Queer Politics and Social Theory*, ed. Michael Warner, 300–320. Minneapolis: University of Minnesota Press.

Cryle, Peter, and Alison Moore. 2011. *Frigidity: An Intellectual History*. London: Palgrave Macmillan.

D'Emilio, John. 1983. *Sexual Politics, Sexual Communities: The Making of a Homosexual Minority in the United States, 1940–1970*. Chicago: University of Chicago Press.

Davidson, Arnold I. 2001. *The Emergence of Sexuality: Historical Epistemology and the Formation of Concepts.* Cambridge, MA: Harvard University Press.
De Bienville, M.D.T. 1886 [1771]. *La nymphomanie, ou traité de la fureur utérine.* Paris: Office de Librairie.
De Lauretis, Teresa. 1991. *Queer Theory: Lesbian and Gay Sexualities.* Bloomington: Indiana University Press.
Dean-Jones, Lesley. 1992. The Politics of Pleasure: Female Sexual Appetite in the Hippocratic Corpus. In *Discourses of Sexuality: From Aristotle to AIDS*, ed. Domna C. Stanton, 48–76. Ann Arbor: University of Michigan Press.
Didi-Huberman, Georges. 2003. *Invention of Hysteria: Charcot and the Photographic Iconography of the Salpêtrière.* Trans. Alisa Hartz. Cambridge, MA: MIT Press.
Downing, Lisa. 2008. *The Cambridge Introduction to Michel Foucault.* Cambridge: Cambridge University Press.
Drescher, Jack, and Joseph P. Merlino, eds. 2007. *American Psychiatry and Homosexuality: An Oral History.* New York: Harrington Park Press.
Epstein, Steven. 1987. Gay Politics, Ethnic Identity: The Limits of Social Constructionism. *Socialist Review* 93/94: 9–54.
Eribon, Didier. 2001. Michel Foucault's Histories of Sexuality. *GLQ: A Journal of Gay and Lesbian Studies* 7 (1): 31–86.
Floyd, Kevin. 2001. Rethinking Reification: Marcuse, Psychoanalysis, and Gay Liberation. *Social Text* 19 (1): 103–126.
Foucault, Michel. 1977. Nietzsche, Genealogy, History. In *Language, Counter-Memory, Practice: Selected Essays and Interviews by Michel Foucault*, ed. Donald F. Bouchard, 139–164. New York: Cornell University Press.
———. 1978. *The Will to Knowledge: The History of Sexuality, Volume 1.* Trans. Robert Hurley. London: Penguin Books.
———. 1980 [1976]. *Power/Knowledge: Selected Interviews and Other Writings, 1972–1977.* New York: Vintage Books.
———. 1985 [1984]. *The Use of Pleasure: History of Sexuality, Volume 2.* Trans. Robert Hurley. New York: Vintage Books.
———. 1991 [1977]. *Discipline and Punish: The Birth of the Prison.* Trans. Alan Sheridan. Middlesex: Penguin Books.
———. 1994 [1977]. Le jeu de Michel Foucault. In *Michel Foucault: Dits et écrits, 1954–1988, tome III 1976–1979*, 298–329. Paris: Gallimard.
———. 1996. The Concern for Truth. In *Foucault Live (Interviews, 1961–1984)*, ed. Sylvère Lotringer, trans. Lysa Hochroth and John Johnston, 455–464. New York: Semiotext(e).
———. 1997a. Polemics, Politics, Problematizations: An Interview with Michel Foucault. In *Ethics: Essential Works of Foucault 1954–1984, Volume 1*, ed. Paul Rabinow, 111–119. London: Penguin Books.
———. 1997b. Sex, Power, and the Politics of Identity. In *Ethics: Essential Works of Foucault 1954–1984, Volume 1*, ed. Paul Rabinow, 163–173. London: Penguin Books.

———. 2011 [1978]. The Gay Science. *Critical Inquiry* 37: 385–403.
Freccero, Carla. 2006. *Queer/Early/Modern*. Durham: Duke University Press.
Freud, Sigmund. 2001 [1905]. *The Standard Edition of the Complete Psychological Works of Sigmund Freud, Volume VII*. Trans. James Stratchey. London: Vintage Books.
Gagnon, John H., and William Simon. 2005 [1973]. *Sexual Conduct: The Social Sources of Human Sexuality*. New Brunswick: Aldine Transaction.
Garton, Stephen. 2004. *Histories of Sexuality: Antiquity to Sexual Revolution*. London: Routledge.
Germon, Jennifer. 2009. *Gender: A Genealogy of an Idea*. New York: Palgrave Macmillan.
Goldstein, Jan, ed. 1994. *Foucault and the Writing of History*. Oxford: Blackwell.
Groneman, Carol. 1994. Nymphomania: The Historical Construction of Female Sexuality. *Signs* 19 (2): 337–367.
Halperin, David. 1990. *One Hundred Years of Homosexuality and Other Essays on Greek Love*. New York: Routledge.
Hekma, Gert. 1991. A History of Sexology: Social and Historical Aspects of Sexuality. In *From Sappho to de Sade: Moments in the History of Sexuality*, ed. Jan Bremmer, 173–193. New York: Routledge.
Huffer, Lynne. 2010. *Mad for Foucault: Rethinking the Foundations of Queer Theory*. New York: Columbia University Press.
Irigaray, Luce. 1977. *Ce sexe qui n'en est pas un*. Paris: Éditions de Minuit.
Jagose, Annamarie. 1996. *Queer Theory*. Melbourne: Melbourne University Press.
Johnson, Mark. 2005. Sexuality. In *Cultural Geography: A Critical Dictionary of Key Concepts*, eds. David Atkinson, Peter Jackson, David Sibley, and Neil Washbourne, 122–127. London: I.B. Tauris.
Katz, Jonathan Ned. 1995. *The Invention of Heterosexuality*. New York: Dutton.
Kent, Kathryn R. 2003. *Making Girls into Women: American Women's Writing and the Rise of Lesbian Identity*. Durham: Duke University Press.
Lotringer, Sylvère. 1988. *Overexposed: Perverting Perversions*. New York: Semiotext(e).
Mauss, Marcel. 1973. Techniques of the Body. *Economy & Society* 2 (1): 70–88.
Maynard, Steven. 1999. 'Respect Your Elders, Know Your Past': History and the Queer Theorists. *Radical History Review* 75: 56–78.
Mazzoni, Cristina. 1996. *Saint Hysteria: Neurosis, Mysticism, and Gender in European Culture*. New York: Cornell University Press.
McCabe, Susan. 2005. To Be and to Have: The Rise of Queer Historicism (Book review). *GLQ: A Journal of Lesbian & Gay Studies* 11 (1): 119–134.
McIntosh, Mary. 1978. The Homosexual Role. *Social Problems* 16 (2): 182–192.
McLaren, Angus. 2007. *Impotence: A Cultural History*. Chicago: University of Chicago Press.

Muñoz, José Esteban. 1999. *Disidentifications: Queers of Color and the Performance of Politics*. Minneapolis: University of Minnesota Press.
O'Leary, Timothy. 2002. *Foucault: The Art of Ethics*. London and New York: Continuum.
Peraldi, François, ed. 1981. *Polysexuality*. New York: Semiotext(e).
Pflugfelder, Gregory M. 1999. *Cartographies of Desire: Male–Male Sexuality in Japanese Discourse, 1600–1950*. Berkeley: University of California Press.
Plummer, Ken. 1973. Awareness of Homosexuality. In *Contemporary Social Problems in Britain*, eds. Roy Bailey and Jock Young, 103–125. Hants and Lexington, MA: Saxon Books and Lexington Books.
———. 1975. *Sexual Stigma: An Interactionist Account*. London: Routledge & Kegan Paul.
Porter, Roy. 1987. *A Social History of Madness*. London: Weidenfeld and Nicolson.
———. 1994. The Literature of Sexual Advice Before 1800. In *Sexual Knowledge, Sexual Science*, eds. Roy Porter and Mikuláš Teich, 134–157. Cambridge: Cambridge University Press.
Reay, Barry, Nina Attwood, and Claire Gooder. 2015. *Sex Addiction: A Critical History*. Cambridge: Polity.
Rich, Adrienne. 1993 [1980]. Compulsory Heterosexuality and Lesbian Existence. In *The Lesbian and Gay Studies Reader*, eds. Henry Abelove, Michèle Aina Barale, and David M. Halperin, 227–254. New York: Routledge.
Rocha, Leon Antonio. 2011. Scientia Sexualis Versus Ars Erotica: Foucault, van Gulik, Needham. *Studies in History and Philosophy of Biological and Biomedical Sciences* 42: 328–343.
Rose, Nikolas. 1998. *Inventing Our Selves: Psychology, Power, and Personhood*. Cambridge: Cambridge University Press.
Sedgwick, Eve Kosofsky. 1990. *Epistemology of the Closet*. Berkeley: University of California Press.
Seidman, Steven, ed. 1996. *Queer Theory/Sociology*. Malden, MA: Blackwell.
Simon, William, and John H. Gagnon. 1967. Homosexuality: The Formulation of a Sociological Perspective. *Journal of Health and Social Behavior* 8 (3): 177–185.
Stein, Marc. 2012. *Rethinking the Gay and Lesbian Movement*. Cambridge: Routledge.
Traub, Valerie. 2008. The Past Is a Foreign Country? The Times and Spaces of Islamicate Studies. In *Islamicate Sexualities: Translations Across Temporal Geographies of Desire*, eds. Kathryn Babayan and Afsaneh Najmabadi, 1–40. Cambridge, MA: Harvard University Press.
———. 2013. The New Unhistoricism in Queer Studies. *PMLA* 128 (1): 21–37.
Turner, William B. 2000. *Genealogy of Queer Theory*. Philadelphia: Temple University Press.

Verhoeven, Tim. 2015. Pathologizing Male Desire: Satyriasis, Masculinity and Modern Civilization at the Fin de Siècle. *Journal of the History of Sexuality* 24 (1): 25–45.

Warner, Michael. 1993. Introduction. In *Fear of a Queer Planet: Queer Politics and Social Theory*, ed. Michael Warner, vii–xxxi. Minneapolis: University of Minnesota Press.

———, ed. 1993. *Fear of a Queer Planet: Queer Politics and Social Theory*. Minneapolis: University of Minnesota Press.

Weed, Elizabeth, and Naomi Schor, eds. 1997. *Feminism Meets Queer Theory*. Bloomington: Indiana University Press.

Weeks, Jeffrey. 1977. *Coming Out: Homosexual Politics from the Nineteenth Century to Present*. London: Quartet Books.

———. 2016. *What is Sexual History?* Cambridge: Polity.

Wittig, Monique. 1992. *The Straight Mind and Other Essays*. Boston: Beacon Press.

CHAPTER 2

Scientia Sexualis and the Patient Case History

The scientific study of sex in the nineteenth century hinged upon discourses that normalised the reproduction of the species. Early sexological treatises proclaimed procreation as the fundamental aim of human existence. For the German-Austrian alienist Richard von Krafft-Ebing, the "propagation of [the] human race" was "guaranteed by the hidden laws of nature which are enforced by a mighty irresistible impulse."[1] Procreation and sexual activity were inextricably entangled such that all sexual acts with non-procreative aims became "a set of symptoms located on a continuum between normality and pathology."[2] In this discursive framework, one's genital composition was taken to determine the direction of sexual feelings, whereby "anatomy equalled psychology, sex physiology determined the sex of feelings."[3] It was also assumed that desire towards one's own sex constituted a form of sexual inversion.[4] For Albert Moll, the impulse that "normal" men experience for women was "natural,"[5] while Iwan Bloch argued that "sexual love constitutes a part of the very being of

[1] Richard von Krafft-Ebing, *Psychopathia Sexualis: A Medico-Forensic Study*, trans. Harry E Wedeck (New York: G. P. Putnam's Sons, 1965 [1886]), 29. "Alienist" is a nineteenth-century term for psychiatrist.

[2] Jennifer Germon, *Gender: A Genealogy of an Idea* (New York: Palgrave, 2009), 131.

[3] Jonathan Ned Katz, *The Invention of Heterosexuality* (New York: Dutton, 1995), 52.

[4] In addition to "sexual inversion", homosexuality was also known as "contrary sexual instinct" and "anthipathic sexual instinct."

[5] Albert Moll, *Les perversions de l'instinct génital: Étude sur l'inversion sexuelle*, trans. Dr Pactet (Paris: Georges Carré, 1893), 234–235.

J. Flore, *A Genealogy of Appetite in the Sexual Sciences*,
https://doi.org/10.1007/978-3-030-39423-3_2

the civilized man."[6] Sexological knowledge was disseminated through publications, conferences and professional research organisations, such as the Scientific Humanitarian Committee, founded by Magnus Hirschfeld and others in 1897, and the *Institut für Sexualwissenchaft* (Institute of Sex Research), which opened in 1919 and was headed by Hirschfeld.

Scientia sexualis, the scientific study of sexuality, emerged prominently in the nineteenth-century writings of central European and American physicians. Case histories of patients were essential to the development of the scientific study of sex and were used as the "raw material" of sexological speculation. From those cases, physicians developed theories, hypotheses, nomenclature and systems of classification that continue to influence contemporary medical thinking on sexuality and gender. This methodology gave rise to what Foucault has called the "*specification of individuals*."[7] The compilation of narratives of patients enabled sexology to present concrete evidence for the existence of psychosexual aberrations.[8] This style of reasoning relies on the interpretation of both the patient's vocalised accounts of themselves and the physician's clinical judgement of those narratives. The patient's history constitutes a document for possible use,[9]

[6] Iwan Bloch, *The Sexual Life of Our Time in its Relations to Modern Civilization*, trans. M. Eden Paul (London: Rebman Limited, 1909 [1906]), 4.

[7] Michel Foucault, *The Will to Knowledge: The History of Sexuality, Volume 1*, trans. Robert Hurley (London: Penguin Books, 1978), 42–43 (emphasis original).

[8] For studies on the development of the patient case history in medicine and the history of sexuality, see Carol Berkenkotter, *Patient Tales: Case Histories and the Uses of Narrative in Psychiatry* (Columbia, SC: University of South Carolina Press, 2008), Ivan Crozier, "Pillow Talk: Credibility, Trust and the Sexological Case History," *History of Science* 46, no. 154 (2008): 375–404, Jonathan Gillis, "The History of the Patient History Since 1850," *Bulletin of the History of Medicine* 80, no. 3 (2006): 490–512, Kathryn Montgomery Hunter, *Doctor's Stories: The Narrative Structure of Medical Knowledge* (Princeton: Princeton University Press, 1991), Thomas Laqueur, "Bodies, Details and the Humanitarian Narrative," in *The New Cultural History*, ed. Lynn Hunt (Berkeley: University of California Press, 1989), 176–204, Harriet Nowell-Smith, "Nineteenth-Century Narrative Case Histories: An Inquiry into the Stylistics and History," *Canadian Bulletin of Medical History* 12 (1995): 47–67, Matt Reed, "*La manie d'écrire*: Psychology, Auto-Observation and Case History," *Journal of the History of Behavioral Sciences* 40, no. 3 (2004): 265–284, Anne Sealey, "The Strange Case of the Freudian Case History: The Role of Long Case Histories in the Development of Psychoanalysis," *History of Human Sciences* 24, no. 1 (2011): 36–50, and John Harley Warner, "The Uses of Patient Records by Historians: Patterns, Possibilities and Perplexities," *Health and History* 1, no. 2–3 (1999): 101–111.

[9] Michel Foucault, *Discipline and Punish: The Birth of the Prison*, trans. Alan Sheridan (London: Penguin, 1991 [1977]), 191.

and it is arranged and produced as a case where individual circumstances are annotated and analysed.[10] The case, Harriet Nowell-Smith writes, thus represents the location where "the theory and practice of medicine converge."[11] The patient case history was essential to accessing the material needed for the expansion of *scientia sexualis*.

Sexual appetite assumed a significant role in the pathologisation of perversions in the nineteenth century. Ideas of balance, frequency and amount functioned to govern sexual normality. Who is "normal" was as much a question of appetite as it was of object choice. This chapter examines how the medicalisation of sexual appetite in nineteenth-century sexology emerged through the technique of the patient case history. It considers two aspects of this technique: first, how the case history presented sexual appetite as a structuring device in the expansion of taxonomies of sexual perversions, and second, how this was accomplished by inextricably tying the imagination and narrative to the notion of sexual excess. The imagination formed the bedrock of sexuality itself and was treated as both essential and suspicious. The patient case history was a discursive device linking pathology, excess and the imagination. It constituted a technique for the ordering of knowledge on sexual appetite and its dissemination.

Locating Appetite in *Psychopathia Sexualis*

In the nineteenth century, patient case histories were presented in the works of early sexologists as evidence for their theories of sexuality.[12] It is this methodology, Ivan Crozier argues, that "makes sexology a medical discipline, rather than simply musing about sexuality."[13] In order to develop taxonomies of sexual deviance, medical writers relied on the words of their patients, letters they received and sometimes analysed published case histories of their contemporaries. The case history chronicled a life, and the life of a body, and in the scene of clinical confession, patients were not only asked to provide facts about "real events," they were also invited to reveal their fantasies, scenarios and dramaturgies. Whether these events had occurred was irrelevant to the pronouncement of pathology. As

[10] Crozier, "Pillow Talk," 378.

[11] Nowell-Smith, "Nineteenth-Century Narrative Case Histories," 50.

[12] Gert Hekma, "'A Female Soul in a Male Body': Sexual Inversion as Gender Inversion in Nineteenth-Century Sexology" in *Third Sex, Third Gender: Beyond Sexual Dimorphism in Culture and History*, ed. Gilbert Herdt (New York: Zone Books, 1993), 217.

[13] Crozier, "Pillow Talk," 376.

Thomas Laqueur has observed, the patient case history shared the techniques of the novel: "[The case] constitutes step-by-step accounts of the history of the body in relation to itself and to social conditions, and provide therefore a model for the intelligibility of misfortune."[14]

Foucault identifies the technique of the case as the "entry of the individual … into the field of knowledge."[15] The case history, buttressed by the anamnesis, became embedded "into the general functioning of scientific discourse."[16] Recast in the language of science, the fictitious characters, performances and practices became expressions of truth. In analysing and classifying those narratives and revelations, the physician also created the picture of an ideal sexuality. In other words, as this chapter goes on to demonstrate, an idealised amount of sexual appetite functioned in the case history in the nineteenth century. The patient case history constituted a technique for the pathologisation and management of sexual appetite. In addition, while physicians, in their discourse of psychiatric illnesses, exercised a form of regulatory power on their patients, as Harry Oosterhuis demonstrates, patients often expressed themselves in contradiction to ideas of "deplorable medical colonization."[17] Individuals thus also

[14] Laqueur, "Bodies, Details, and the Humanitarian Narrative," 181–182.

[15] Foucault, *Discipline and Punish*, 191.

[16] Ibid.

[17] Harry Oosterhuis, *Stepchildren of Nature: Krafft-Ebing, Psychiatry and the Making of Sexual Identity* (Chicago: University of Chicago Press, 2000), 10. It is worth noting that Karl Heinrich Ulrichs (1825–1895) and Karl-Maria Benkert (1824–1882), who both contributed to nineteenth-century writings on homosexuality and made efforts at classification while campaigning for reform, were not trained in medicine. Ulrichs introduced *uranism* (homosexuality) in 1864, while the writer Karl-Maria Benkert coined "homosexuality" in 1869; both labels "were actually of a nonmedical proto-emancipatory origin." Oosterhuis, *Stepchildren of Nature*, 44. A key political issue in late-nineteenth-century Germany was the reform of Paragraph 175, a provision of the criminal code which criminalised what it referred to as "unnatural vice," that is, sex between men and bestiality. See further, Harry Oosterhuis, "Albert Moll's Ambivalence about Homosexuality and His Marginalization as a Sexual Pioneer," *Journal of the History of Sexuality* 28, no. 1 (2019): 1–43, Tracie Matysik, "In the Name of the Law: The 'Female Homosexual' and the Criminal Code in Fin de Siecle Germany." *Journal of the History of Sexuality* 13, no. 1 (2004): 26–48, Karl Heinrich Ulrichs, *Riddle of Man-Manly Love: The Pioneering Work on Male Homosexuality*, trans. Michael A. Lombardi-Nash (Buffalo: Prometheus Books, 1994), Hubert Kennedy, *Ulrichs: The Life and Works of Karl Heinrich Ulrichs, Pioneer of the Modern Gay Movement* (Boston: Alyson Publications, 1988), Manfred Herzer, "Kertbeny and the Nameless Love," *Journal of Homosexuality* 12, no. 1 (1986): 1–26, and Judit Takács, "The Double Life of Kertbeny," in *Past and Present of Radical Sexuality Politics*, ed. Gert Hekma (Mosse Foundation: Amsterdam, 2004), 26–40.

performed or enacted an ethics of personhood and sexuality, and manifested their "identity."

In *The Will to Knowledge*, Foucault explains how the practice of confession became a central component of *scientia sexualis* and he locates it within a production of discourses of truth about sexuality. "The obtaining of the confession and its effects," he writes, "were recodified as therapeutic operations."[18] In the patient case history, the physician's judgement, and the patient's confession, combined in the normalisation of psychosexual disorders. This "narrative-within-a-narrative"[19] operated in a field of sexological objects from which it was inextricable. Sexologists harnessed the patient case history in various ways, sometimes reproducing the patient's own words, and at other times, physicians edited anamneses and added their clinical judgement. Albert von Schrenck-Notzing, a German psychiatrist perhaps best known for using hypnotism as therapy and espousing "para psychology," for instance, made this aim explicit in *Therapeutic Suggestions*: "for the sake of objectivity, I have allowed patients to speak for themselves, and give their letters verbatim."[20] Addressing doubt in the veracity of patients' words, Schrenck-Notzing also wrote that it was up to readers to form their own judgement "of the conscientiousness and objectivity of the descriptions, from the histories of the cases."[21]

Schrenck-Notzing's publication of case studies represents both the reality of the patient's condition and authenticity of disease. The case history is the basic unit, the foundational material of medical thought and, however scientific the case may be, it is organised, presented and disseminated as a narrative.[22] Located in a practice of writing and speaking, the patient's story becomes part of a "diagnostic plot."[23] The gathering and compilation of narratives of patients allowed sexology to present concrete evidence for the existence of psychosexual aberrations. In the writing of the medical text, physicians availed themselves of the individual case

[18] Foucault, *The Will to Knowledge*, 67.

[19] Berkenkotter, *Patient Tales*, 2.

[20] Albert von Schrenck-Notzing, *Therapeutic Suggestions in Psychopathia Sexualis with Especial Reference to Contrary Sexual Instinct*, trans. Charles Gilbert Chaddock (Philadelphia: The F. A. Davis Company, 1895), viii. See further Heather Wolffram, *The Stepchildren of Science: Psychical Research and Parapsychology in Germany, c. 1870–1939* (Amsterdam: Rodopi B.V., 2009).

[21] Schrenck-Notzing, *Therapeutic Suggestions*, viii.

[22] Hunter, *Doctor's Stories*, 51.

[23] Ibid., 131.

history to expound on other related conditions and formulated medical and scientific conclusions on those.[24] From individual circumstances, then, is produced a case where behaviours can be annotated, analysed and categorised.

The kind of cases that collectively informed the emergence of a science of sex included the confessions of patients who spoke not only of their desires and fantasies, but also their family history. Nearly all cases compiled by Krafft-Ebing for instance commenced with an outline of diseases found in the patient's parentage. This concern with the heredity of sexual perversions and other conditions, such as neurasthenia, was characteristic of the popularity of the theory of degeneration in the nineteenth century. Degeneration was posited by Bénédict Augustin Morel's *Traité des dégénérescences physiques* (1857) in which the French psychiatrist emphasised the congenital quality of degeneration and racial deterioration as well as the effects of the environment on this purportedly retrogressive state.[25] In an example of how case studies were presented, Morel recorded the familial lineage of "cretinised" (*crétinisé*) individuals.

Morel was a key figure of the theory of generation in mid-nineteenth-century France; in Germany, on the other hand, the physician Max Simon Nordau denounced the over-stimulation of the imagination in his diagnoses of degeneration. He asserted that "unconventional" artists of the *fin-de-siècle* were degenerates who indulged in luxury and were afflicted with excessive sensibilities.[26] As Daniel Pick observes, degeneration functioned

[24] The publication and analysis of case histories were often challenged by sexologists, who would then publish revised interpretations. This practice made the sexual sciences (and its taxonomies) a uniquely dynamic domain of study in the late-nineteenth century. See further Heike Bauer, *English Literary Sexology: Translations of Inversion 1860–1930* (Basingstoke: Palgrave Macmillan, 2009) and *Sexology and Translation: Cultural and Scientific Encounters Across the Modern World* (Philadelphia: Temple University Press, 2015), Crozier, "Pillow Talk" and Oosterhuis, *Stepchildren of Nature*.

[25] Bénédict Augustin Morel, *Traité des dégénérescences physiques, intellectuelles et morales de l'espèce humaine et des causes qui produisent ces variétés maladives* (Paris: J. B. Baillière, 1857).

[26] Nordau formulated interesting ideas on the imagination and the stimulation of the mind in the chapter on mysticism, which he called the "cardinal mark of degeneration." Max Simon Nordau, *Degeneration* (New York: D Appleton & Company, 1895 [1892]), 22. See also 60–66.

as "the condition of conditions, the ultimate signifier of pathology."[27] Degeneration operated within a similar framework as social Darwinism, which was popular by the late 1870s. By the end of the nineteenth century, socio-biological theories widely influenced understandings of sexuality and sexual deviance. This was linked to anxiety about nationalism and social hygiene; the "survival of the fittest necessitated a healthy national organism, free of hereditary disease and moral weakness."[28] Degeneration was harnessed to analyse patients' sexual lives, and conditions such as sexual excess and sexual lack were examined as manifestations of corrupted or devitalised sexual appetite.

Historians of sexuality have largely chronicled the emergence of sexology in the nineteenth century as an instance of the medicalisation of homosexuality and sexual object choice more broadly. In his account of the regulation of sexuality, Weeks argues that Krafft-Ebing conceived of a "firmly heterosexual drive" which "became the orthodox view in the late nineteenth century."[29] However, this "drive" in Krafft-Ebing's works was also contingent upon ideas of appetite and balance. The drive of which Krafft-Ebing spoke at length was not solely a question of object choice, as it could result, for example, in complete impotence. An understanding of drive as subject to amount, degree and balance circulates in the work of Krafft-Ebing and his contemporaries. Indeed, the patient case history acts a conduit for this conceptualisation of the problem of sexual appetite.

[27] Daniel Pick, *Faces of Degeneration: A European Disorder, c.1848–1918* (Cambridge: Cambridge University Press, 1989), 8. Several scholars have noted that harnessing degeneration offered considerable advantages to psychiatrists. Ian R. Dowbiggin observes that degeneration solved several professional difficulties and served to expand the terrain of psychiatric practice. It enabled the field to gain scientific legitimacy since conclusive somatic proof of mental insanity was still unsubstantiated. As Oosterhuis further writes, "It is difficult to escape the impression that psychiatrists consciously or unconsciously capitalized on the imprecision of degeneration theory in order to divert attention away from the lack of empirical evidence of the somatic basis of mental illness." Oosterhuis, *Stepchildren of Nature*, 106–107. See also Jan Goldstein, *Console and Classify: The French Psychiatric Profession in the Nineteenth Century* (Cambridge: Cambridge University Press, 1981) and Ian R. Dowbiggin, *Inheriting Madness: Professionalization and Psychiatric Knowledge in Nineteenth-century France* (Berkeley: University of California Press, 1991).

[28] George L. Mosse, "Nationalism and Sexuality in Nineteenth-Century Europe," *Culture & Society* 20 (1983): 78.

[29] Jeffrey Weeks, *Sex, Politics and Society: The Regulations of Sexuality Since 1800* (London: Routledge, 2012 [1981]), 4. See also Jörg Hutter, "The Social Constructions of Homosexuals in the Nineteenth Century: The Shift from the Sin to the Influence of Medicine in Criminalizing Sodomy in Germany," *Journal of Homosexuality* 24, no. 3–4 (1993): 73–93.

In his history of heterosexuality, Katz employs the case of Z to argue that Krafft-Ebing's use of "homosexual" and "heterosexual" "helped to make sex difference and eros the basic distinguishing features of a new linguistic, conceptual, and social ordering of desire."[30] Desire here is approached in terms of *direction*, rather than *degrees*. Yet, in the case study quoted by Katz, Krafft-Ebing identified *amount* as a key issue. Katz considers the case study of Z as demonstrating how psychiatrists would subject "clients inclined to both sexes to especially severe moral censure when they continued, wilfully, to follow their same-sex desires."[31] Psychiatrists certainly sought to "correct" the direction of their patients' desires, for example, by advising so-called inverts to visit prostitutes. However, Krafft-Ebing, in his conceptualisation of such sexual inclinations, also considered their *strength*, that is, their amount, as highly significant. In his case notes on Z, Krafft-Ebing wrote:

> The main object was to strengthen the sexual inclination for the opposite sex, which was defective, but not absolutely wanting. This could be done by … the excitation and exercise of normal sexual desires and impulses.[32]

Hence, Krafft-Ebing thought of same-sex desires as inextricable from notions of force or strength. The reference to "inclination" further suggests that Krafft-Ebing thought of Z's appetite as requiring balance, a sort of stabilisation of his urges, towards the opposite sex.

In Krafft-Ebing's *Psychopathia Sexualis* (1886), Z's case is found under "Antipathic Sexuality," a term used to describe sexual inversion. The direction of Z's attraction is problematised, and hence classified under this category, alongside his sexual appetite. In Krafft-Ebing's opus, we find multiple cases outlining "degrees" of inverted sexual feelings, some of them "acquired" and others "inborn."[33] Under the classification of sexual inversion, Krafft-Ebing worked to further group individuals according to *degrees*. Krafft-Ebing, and other sexologists at the time, viewed amount as *inextricable* from object choice and considered sexual inversion itself as a matter of degree.

[30] Katz, *The Invention of Heterosexuality*, 28.
[31] Ibid., 24
[32] Krafft-Ebing, *Psychopathia Sexualis*, 377–378.
[33] Ibid., 313 and 320.

The taxonomy of sexual perversions was fundamental to the expansion of sexology in the nineteenth century. In several sexological treatises, writers proceeded to classify a range of sexual behaviours as perversions. Krafft-Ebing's *Psychopathia Sexualis*, drawing extensively on case notes and letters, deployed a four-part classification: *paradoxia* (wrong time, too young or too old), *anaesthesia* (wrong amount, too little), *hyperaesthesia* (wrong amount, too much) and *paresthesia* (wrong object choice). This treatise has been widely read in terms of its understanding of *paresthesia*. Hekma, for instance, argues that by the 1880s the classification systems were based on biological understandings of sexual perversions that "replaced older notions of exhaustion, lust, and excessive fantasy."[34] This comment is echoed by Ladelle McWhorter who notes that while a range of "sexual subjectivities" such as "zoophiles and zooerast … have passed away … the homosexual has remained and helped give shape to his 'normal' counterpart, the heterosexual."[35] Conversely, I suggest that "normal" is not confined to objects, and ideas of lust, excess and lack circulate in the work of Krafft-Ebing as a matter of degree between normal and abnormal.[36] Further, as Peter Cryle and Elizabeth Stephens note, the cases reveal a "dynamic and unstable" relationship between normal and abnormal.[37]

Although Krafft-Ebing has been widely associated with the historical pathologisation of homosexuality, he considered sadism and masochism as fundamental aberrations. They represented polar opposites of the sexual instinct and "such disorders were also frequently combined … with other forms of sexual perversion."[38] Sadism and masochism were paradigms of perversions because "with these two disorders sexual hyperesthesia reaches its most bizarre extreme."[39] Sexual hyperesthesia, or sexual excess, was of particular concern for Krafft-Ebing. The focus on the highest degree of

[34] Gert Hekma, "A History of Sexology: Social and Historical Aspects of Sexuality," in *From Sappho to de Sade: Moments in the History of Sexuality*, ed. Jan Bremmer (New York: Routledge, 1991), 180.

[35] Ladelle McWhorter, *Bodies and Pleasures: Foucault and the Politics of Normalization* (Bloomington: Indiana University Press, 1999), 32.

[36] Peter Cryle and Elizabeth Stephens, *Normality: A Critical Genealogy* (Chicago: University of Chicago Press, 2017), 270.

[37] Ibid., 274.

[38] Frank J. Sulloway, *Freud, Biologist of the Mind: Beyond the Psychoanalytic Legend* (New York: Basic Books, 1979), 294.

[39] Ibid., 287.

sexual perversion reflects the problem of sexual appetite and its varying amounts. Alison Moore observes that Krafft-Ebing also conceived of sadistic and masochistic activity in terms of their degree, "Minimal sadism was understood to be the normal predilection of heterosexual men in civilized European society. Excess sadism, on the other hand, was barbarous, and hence its manifestation in the modern era a sign of retrograde degeneration."[40]

Appetite was internal to Krafft-Ebing's classification of sadism and masochism. Not only did he believe some form of sadism to be normal in men (and some form of masochism to be normal in women), he also considered that such appetites in excess were pathological. Sadism and masochism appear in Krafft-Ebing's case studies connected to broader notions of appetite. In addition, Krafft-Ebing argued that sexual hyperaesthesia "must always be regarded as the basis of sadistic inclinations. The impotence which occurs so frequently in psychopathic and neuropathic individuals [results] from excesses practiced in early youth."[41] The patients' cases became devices for positioning sadism in terms of sexual hyperaesthesia, that is, in terms of its excessive amounts and its inexorable consequence, impotence.

The patient's case history constituted a dynamic discursive space where physicians could develop their ideas on sexual appetite, while also revising the works of their contemporaries. Julia Epstein writes that in the "clinical case record, language mediates bodily experience so that such experience can be made available for interpretation."[42] Cases were reinterpreted in different frameworks or different systems of classification of sexual appetite. Schrenck-Notzing, in a treatise on sexual inversion, modified Krafft-Ebing's classification of *hyperaesthesia* and divided it into two "classes": the first, onanism (auto-sexual indulgence) and the second, satyriasis and nymphomania (which he also called "hetero-sexual indulgence").[43] Sexual excess and sexual lack were both central and peripheral to sexual inversion. Schrenck-Notzing wrote that onanism and impotence were "constant accompaniments of sexual perversions" that are key to "understanding of

[40] Alison Moore, "The Invention of Sadism? The Limits of Neologisms in the History of Sexuality," *Sexualities* 12, no. 4 (2009): 487.

[41] Krafft-Ebing, *Psychopathia Sexualis*, 116.

[42] Julia Epstein, "Historiography, Diagnosis, and Poetics," *Literature and Medicine* 11, no. 1 (1992): 38.

[43] Schrenck-Notzing, *Therapeutic Suggestions*, 2.

paresthesia sexualis."[44] Their presence designated a pathological condition, but their study was also subordinated to that of the contrary sexual instinct.

Schrenck-Notzing's discussions of sexual inversion contained multiple references to intensity and impulse. The concept of sexual appetite was harnessed to organise knowledge on sexual perversions in his writings. He, for instance, conceived of direction and intensity of sexual appetite as inseparable: "abstinence from natural sexual congress induces sexual hyperesthesia and intensifies the impulse in its perverse direction."[45] In the case history of his patient K, who Schrenck-Notzing treated using hypnotism, the physician noted that the patient enjoyed frequent intercourse with prostitutes before "growing aversion to the female sex." K then "gave up his commission as an officer because soldiers excited him so sexually."[46] During his fourth visit, K wished "to try his luck with women but fears impotence," and in the final consultation, he felt "as little interest in women as in men."[47] In Schrenck-Notzing's case notes of the patient's history, sexual appetite fluctuates and co-exists alongside sexual object choice. Contributions to systems of classification did not result in a precise, coherent taxonomy of sexual aberrations. Nonetheless, it revealed the medical attention that was devoted to fluctuations of sexual appetite.

Journeys in Fetishism: The French Intervention

Nineteenth-century efforts to develop nomenclature on sexual perversions were by no means homogeneous. Nonetheless, whether focused on sexual inversion, or sadism and masochism, physicians continually returned to the problem of sexual appetite. Through the technique of the patient case history, appetite was integral to the development of nosologies of sexual kinds and to the expansion of a science of sex. While French physicians at the turn of the century usually classified "all perversions under a single nosological entity,"[48] in particular, sexual inversion and fetishism, they also provided a language through which the ebbs and flows of sexual appetite became problematised. In 1887, French psychologist Alfred

[44] Ibid., v.
[45] Ibid., 207.
[46] Ibid., 215.
[47] Ibid.
[48] Anna Katharina Schaffner, *Modernism and Perversion: Sexual Deviance in Sexology and Literature, 1850–1930* (Basingstoke: Palgrave Macmillan, 2012), 65.

Binet introduced the term "fetishism" to medical nomenclature. For Binet, fetishism was a form of pathological and sexual fixation on an object, a body part or a feature of someone's appearance (e.g., a fetish for short-haired women). In his concluding chapter on defining fetishism, Binet declared that what distinguishes fetishism from "normal love" is a question of *degree*: "we can say that [fetishism] exists in a latent form in normal love; the germ only needs to blossom for perversion to appear."[49] Fetishism was an *idée fixe*; a desire that dominates all thought and distorts "normal" desires. Fetishism was an all-encompassing disease, which could manifest itself through different objects of desire, whether human or otherwise. "Normal love," as Binet phrased it, always carried the dangers of certain excesses. Fetishistic perversions, Robert Nye writes, "were a product of weakened procreative activity."[50] The *vita sexualis* needed equilibrium, as was the case with the nineteenth-century human body. Hence, as Nye argues, "Moderate (average) rates of expenditure were deemed normal, and excessive or insufficient rates were judged to fall in the range of the pathological."[51] The sexological economy of "normal" and "pathological" was a question of intensity as much as one of object choice—physicians continued to problematise *how much* sexual appetite was suitable in individuals. Normality emerged in physicians' writings through the question of degree and intensity.[52]

Interest in fetishism has been unpacked by way of geographical and cultural issues. Nye has explained the French interest in fetishism in the late nineteenth century as representing a confluence of national and demographic concerns hinging on sterility and declining population growth.[53] Schaffner, on the other hand, suggests that fetishism became a key perversion for Binet due to his reliance on literary texts. Binet for instance read Jean-Jacques Rousseau's *Confessions* as a case study in fetishism.[54] While the importance of fictional texts of literature in the works of several physicians, including Binet and Krafft-Ebing, certainly revealed their approach

[49] Alfred Binet, *Le fétichisme dans l'amour* (Paris: Octave DOIN, 1888), 272.

[50] Robert A. Nye, "The History of Sexuality in Context: National Sexological Traditions," *Science in Context* 4, no. 2 (1991): 399. See also Robert A. Nye, "The Medical Origins of Fetishism" in *Fetishism as Cultural Discourse*, eds. Emily Apter and William Pietz (Ithaca and London: Cornell University Press, 1993), 13–30.

[51] Nye, "The Medical Origins of Fetishism," 16.

[52] See Cryle and Stephens, *Normality*.

[53] Nye, "The History of Sexuality in Context."

[54] Schaffner, *Modernism and Perversion*, 80–81.

to sexual perversions and the function of literature in their texts, I propose that the focus on fetishism in Binet's works, as well as sadism and masochism in the work of Krafft-Ebing, demonstrates an attention to the importance of amount and degrees of sexual appetite, for which the patient case history was a vehicle. Binet cited the case of Mr X and argued that for the fetishist, the sensory perception of the love-object created a pleasure more intense than the sexual feeling.[55] The case history of X indicates the issues of sensuality and the intensity of pleasure. The taxonomies of sexual perversions deployed by the physicians, while often inconsistent, demonstrate a clear concern with how sexual appetite acts as a "strain" on individuals such that they are pushed towards, or away from, sexual activity. In fetishism, sexual appetite "fixates" individuals on certain people or objects.

The idea of obsessive fixations was taken up by Valentin Magnan, a psychiatrist and frequent collaborator of Jean-Martin Charcot,[56] although he designated sexual inversion as the principal perversion. However, much like Binet and others, Magnan devoted medical attention to the question of intensity. In *Des anomalies, des aberrations et des perversions sexuelles* (1867), Magnan developed a system of classification, blending neurological and anatomical symptoms. He detailed cases of inversion, fetishism, nymphomania and satyriasis, and finally obsessive but platonic love as forms of "atavistic evolutionary regression to a pre-civilized state of being."[57] Magnan presented a four-part classification system: spinals (*spinaux*), posterior spinal cerebral (*spinaux cérébraux postérieurs*), anterior spinal cerebral (*spinaux cérébraux anterieurs*) and anterior cerebral or psychic (*cérébraux anterieurs, ou psychiques*). Magnan then devoted a large part of his theories to excessive sexual appetite. The intensity of such appetites could drive individuals to "obsessions, impulsions and various delirium."[58]

In *Les centres nerveux* (1893), Magnan deployed the same taxonomy and again returned to the question of excessive sexual behaviours. He identified a recurrent cause to homicidal behaviour, kleptomania, pyromania and exhibitionism: "The phenomenon is the same everywhere: it is

[55] Binet, *Le fétichisme dans l'amour*, 270.

[56] See Georges Didi-Huberman, *Invention of Hysteria: Charcot and the Photographic Iconography of the Salpêtrière*, trans. Alisa Hartz (Cambridge, MA: MIT Press, 2003).

[57] Schaffner, *Modernism and Perversion*, 67.

[58] Valentin Magnan, *Des anomalies, des aberrations et des perversions sexuelles* (Paris: A. Delahaye & E. Lecrosnier, 1885), 27 (translation author).

always an overexcited centre which interpellates the sensation or the act that will satisfy the urge."[59] It would seem, from Magnan's treatise, that overexcitement, the inexorable pull of a certain urge, emerged from a similar demand, and a similar kind of appetite. This overexcited centre did not necessarily lead to "perversion"; it could lead to criminal and anti-social acts.

Magnan attributed to this pull towards overexcitement the ability to completely incapacitate an individual, rendering any erotic and sensual activity impossible. For Magnan, "an obsession [could] also exert a real paralysing power upon the sexual act."[60] He cited as example the case of a twenty-one-year-old man so transfixed by the number thirteen that it would completely "freeze his virility" ("*glace sa virilité*").[61] Magnan and Binet thus suggested a similar idea about the core of excitement that affected an array of behaviour, sexual and otherwise. While Binet incorporated appetite through the notion of "degree" into a pathologisation of improper objects and body parts, Magnan suggested the existence of an "overexcited centre" that acts as a strain on the individual where she or he is pushed towards aberrations—whether this means excessive sexual activities, or so-called frozen sexuality.

Rhythms of Pleasure

This chapter has so far demonstrated how sexual appetites became implicated in discourses of perversion, whether fetishism or inversion, across different national sexological traditions in the nineteenth century. Appetite required moderation in the construction of the "normal" and balanced sexual subject. In the production of taxonomies and their concomitant circulations, the patient case history constituted an anchor that provided the "proof" sexology needed for its own legitimacy. In the dissemination of discourses on the perversions, sexologists of the late nineteenth century recognised that sexuality was susceptible to flows. Of interest here are Havelock Ellis' ideas on sexual periodicity and *rhythms*. Rhythm for Ellis was all-encompassing: "Rhythms, it is scarcely necessary to remark, is far from characterizing sexual activity alone. It is the character of all biological

[59] Valentin Magnan, *Recherches sur les centres nerveux: Alcoolisme, folie des héréditaires dégénérés, paralysie générale, médecine légale* (Paris: G. Masson, 1893), v (translation author).

[60] Ibid., 166.

[61] Ibid., 167.

activity, alike on the physical and psychic sides."[62] Ellis was also known for claiming that impotence and frigidity were more significant issues than sexual excesses.[63] Like several authors studied in this chapter, Ellis' theories on sexual inversion have been privileged in historical studies, rather than his attention to appetite, balance and rhythm.[64]

Ellis, an English writer and physician, argued that "sexual functions are periodic" and closely connected to menstrual cycles in women.[65] He also posited that monthly cycles might exist in men. One of Ellis' cases concerned a man who claimed that he was "just like a woman, always most excitable at a particular time of the month."[66] Ellis drew on detailed case notes to provide more information on sexual rhythms. One case in particular, taken from the diaries of X, charts the changes in amount of sexual appetite on different dates over 1892–1893. This appeared alongside notes on whether sexual intercourse or nightly emission occurred. For example, "Wednesday, May 3, 1893. The peculiar feeling … had sexual relations, and [then it] disappeared."[67] Ellis' discussion of rhythms and amounts extended over several pages, sometimes outlining the excesses and dwindling of relations. Variations in sexual appetite prominently featured in Ellis' patient case histories, often accompanied by analyses of their connections to menstruation. As Crozier observes, in sexology, "the burden of proof lay squarely with how people actually behaved."[68] The documentation of changes depending on the time of the month and other

[62] Havelock Ellis, *Studies in the Psychology of Sex, Volume 1* (London: William Heinemann Medical Books, 1942 [1905]), 85.

[63] Ibid., 219.

[64] See, for example, Joseph Bristow, "Symonds's History, Ellis's Heredity: *Sexual Inversion*," in *Sexology in Culture: Labelling Bodies and Desires*, eds. Lucy Bland and Laura Doan (Cambridge: Polity Press, 1998), 79–99, Chris Waters, "Havelock Ellis, Sigmund Freud and the State: Discourses of Homosexual Identity in Interwar Britain," in *Sexology in Culture: Labelling Bodies and Desires*, eds. Lucy Bland and Laura Doan (Cambridge: Polity Press, 1998), 165–179, and Ivan Crozier, ed., *Havelock Ellis and John Addington Symonds, Sexual Inversion: A Critical Edition* (Basingstoke: Palgrave Macmillan, 2008). In *Frigidity*, Cryle and Moore analyse Ellis's ideas on frigidity; however, they do not take up sexual periodicity and rhythms. See Peter Cryle and Alison Moore, *Frigidity: An Intellectual History* (London: Palgrave Macmillan, 2011), 212–215.

[65] Ellis, *Studies in the Psychology of Sex, Volume 1*, 85.

[66] Ibid., 112.

[67] Ibid., 119.

[68] Ivan Crozier, "Havelock Ellis, Eonism and the Patient's Discourse; or, Writing a Book about Sex," *History of Psychiatry* 11, no. 42 (2000): 147.

factors needed to be communicated by patients and recorded by physicians in case histories.

The importance of fluctuations in sexual appetite was also embedded within the conceptualisation of the sexual instinct itself. The German psychiatrist Albert Moll originally introduced the term "detumescence" in *Handbuch der Sexualwissenschaften* (*Handbook of Sexologies*), published in 1912. Detumescence was accompanied by *Kontrektationstrieb* (contrectation drive). Contrectation represented a drive to seek intimate physical and emotional connection. Moll's introduction of contrectation-impulse and detumescence-impulse indicates that the sexologist considered sexual appetite key to understanding the sexual impulse more broadly. The concepts of drive and discharge evoke images of energy fluctuations in the framework of sexual appetite. This was similar to Ellis' conceptualisation of the sexual instinct. Ellis contended that all sexual functioning shared the processes of tumescence (arousal) and detumescence (release). Tumescence, Ellis wrote, "is to be found in the fact that vascular congestion, more especially of the parts related to generation, is an essential preliminary to acute sexual desire."[69] He added the tumescence occurred before "desire can become acute,"[70] and that tumescence and detumescence were "fundamental, primitive, and essential" processes of the sexual impulse.[71] Ellis conceded that the release (detumescence) could be achieved through various means including onanism, oral sex, anal sex and vaginal penetration, depending on the case. He spoke of tumescence in terms of degrees enhanced through the senses. Sight and touch were particularly relevant to tumescence and detumescence. The "sensory channels," Ellis argued, act as transmitters of "stimuli," which in turn "exert on the strength and direction of the sexual impulse."[72] The senses, in Ellis' writings, thus affect sexual appetite and are sensitive to fluctuations themselves.

Moll's and Ellis' writings, it should be noted, foreshadowed Sigmund Freud's ideas on the sexual instinct that would eventually sever links between sexual aim and sexual object choice. "It seems probable," Freud posited, "that the sexual instinct is in the first instance independent of its

[69] Ellis, *Studies in the Psychology of Sex, Volume 1*, 25.

[70] Ibid.

[71] Ibid., 27.

[72] Ibid., 2. Ellis also wrote that the "chief stimuli which influence tumescence and thus direct sexual choice come chiefly—indeed, exclusively—through the four senses." Ibid., 1.

object; nor is its origin likely to be due to its object's attractions."[73] Moll's attention to sexual appetite is reflected in *The Sexual Life of the Child* (1909), where he introduced "stages of the voluptuous sensation," an early example of the conceptualisation of the human sexual response in relation to rhythms of pleasure. In this work, Moll drew on a range of material including, unpublished diaries, autobiographies, albums and memoirs to localise sexual excitement throughout the body.[74] Hunter writes that clinical judgement can be thought of as the ability to discern a plot from data assembled from the patient and other sources.[75] In other words, when it comes to sexual function, the physicians must draw from patients' words and clinical readings, to be able to communicate complex patterns and schemas. Moll mobilised cases to formulate four phases of sexual response, within which appetite was crucial: "its onset; the equable voluptuous sensation; the voluptuous acme, coincident with the rhythmical contraction of the perineal muscles and the ejaculation of the semen; and finally, the quite sudden diminution and cessation of the voluptuous sensation."[76]

In the mid-to-late nineteenth century, sexual appetite included much more than the sexual "urge" or sexual "instinct." Sexual appetite operated as a kind of structuring device to an inappropriate object choice and it was harnessed to explain how the sexual instinct could go awry and fixate on improper objects. Indeed, throughout the cases detailed by sexological thinkers, the concern with love-objects is both prevalent and configured through amounts, rhythms and excitement. The emphasis on perversions and the pathologisation of sexual inversion has led historians of sexuality to overlook the importance of dimensions of appetite in sexology. Arnold I. Davidson, for example, considers that in the nineteenth century, sadism, masochism, fetishism and homosexuality all "exhibit the same kind of perverse expression of the sexual instinct, the same basic kind of functional deviation, which manifests itself in the fact that psychological satisfaction

[73] Sigmund Freud, *The Standard Edition of the Complete Psychological Works of Sigmund Freud, Volume VII*, trans. James Strachey (London: Vintage Books, 2001 [1905]), 148. See also 136–148.

[74] Moll, for example, refers to the autobiography of Felix Platter, a sixteenth-century Swiss physician. Albert Moll, *The Sexual Life of the Child*, trans. Eden Paul (New York: The Macmillan Company, 1912 [1909]), 10–11. See 136–141 for more examples of autobiographies.

[75] Hunter, *Doctor's Stories*, 45.

[76] Moll, *Sexual Life of the Child*, 22–23.

is obtained primarily through activities disconnected from the natural function of the instinct."[77]

Questions of functionality of the sexual instinct certainly were circulating in medical treatises in the nineteenth century. However, this chapter suggests that sexual appetite also structured ideas on these disorders and connected them conceptually through the patient case history. Sexologists writing on disorders of sexual instinct tied it to a "normal" amount of sexual appetite. "Wrong" objects were sought because of imbalances in sexual appetites. For instance, Moll asserted that long periods of abstinence in inverts could result in sickness: "When [inverts] appear ill or nervous, it is probably due to forced abstinence, rather than sexual excess."[78] Hence, it would be mistaken to encourage male inverts to abstain from sex with other men in the hope that they would be forced to turn to women. Instead, abstinence would worsen the "morbid state" of inverts.[79] The question of "wrong" sexual object choice was intimately tied to amounts of sexual appetite and total abstinence could lead to morbid sexual excesses. Likewise, as observed above, Binet spoke of "normal" love in terms of degrees, while Krafft-Ebing continually referred to *anaesthesia sexualis* and *hyperaesthesia* throughout his work.

Narrating *Phantasia Morbosa*

In the patient case histories of the nineteenth century, sexual pathologies were associated with misuses of the imagination and the enfeeblement of nerves. This was closely connected to the problem of masturbation. As Laqueur writes,

> The connections between literary practices and masturbation are deep and extensive. Masturbation's evils—suspicious solitude, dependence on the perfervid and unbounded imagination, the seeming inevitability of addictive excess—find parallels in the silent but far-reaching revolution of consciousness that private reading both reflects and helped create.[80]

[77] Arnold I. Davidson, *The Emergence of Sexuality: Historical Epistemology and the Formation of Concepts* (Cambridge, MA: Harvard University Press, 2001), 76.

[78] Moll, *Les perversions de l'instinct génital*, 295 (translation author).

[79] Ibid.

[80] Thomas Laqueur, *Solitary Sex: A Cultural History of Masturbation* (New York: Zone Books, 2003), 306.

Reading patterns, not just the kinds of material read, but also the amount of reading in which patients indulged, were associated at the *fin-de-siècle* with masturbation. Nineteenth-century physicians, echoing eighteenth-century preoccupations with solitude, treated private reading with suspicion. The imagination of the patient, which was utilised by readers and communicated to physicians, was studied to isolate signs of mental ailment. Through the patient's narrative of childhood, family history and current condition, the imagination of the reader was both revealed and used to problematise sexual appetite.

In his lecture series *Abnormal*, Foucault credits a work by Russian physician Heinrich Kaan titled *Psychopathia Sexualis* (1844) as "the first treatise of psychiatry to speak only of sexual psychopathology."[81] Kaan's text was almost entirely devoted to medical and pedagogical warnings against the dangers of onanism. For him, onanism was the primary perversion, followed by "the love of boys … lesbian love, the violation of cadavers, sex with animals, and the satisfaction of lust with statues."[82] In addition, Kaan recognised that the "sexual drive (*Geschlechtstrieb*) displays numerous variations with respect to quantity, likewise it also deviates from a standard norm with respect to quality."[83] Thus, sexual amount was intertwined with "quality" as well as object choice. In arguing this, Kaan produced a concept of normality that encompassed dimensions of appetite, quality and direction.

Onanism was especially concerning because of its connections to an excessive imagination or *phantasia morbosa*. In fact, as Laqueur has pointed out, masturbation was deemed an excessive practice, one that invoked and produced uncontrollable appetite and intractable needs. It was a force that needed control because it is "normally excessive."[84] In his editor's introduction to the first English translation of Kaan's book, Benjamin Kahan argues that Kaan provided sexuality a "new structure—imagination—that links all sexual acts together and that connects bodily instincts to the mind."[85] Kahan expresses the hope that the translation of *Psychopathia Sexualis*, which was originally written in Latin, will "inaugurate a wholesale

[81] Michel Foucault, *Abnormal: Lectures at the Collège de France, 1974–1975*, trans. Graham Burchell (New York: Picador, 2003), 278.

[82] Benjamin Kahan, ed., *Heinrich Kaan's "Psychopathia Sexualis" (1844): A Classic Text in the History of Sexuality*, trans. Melissa Haynes (Ithaca: Cornell University Press, 2016), 78.

[83] Ibid.

[84] Foucault, *Abnormal*, 278.

[85] Kahan, *Heinrich Kaan's "Psychopathia Sexualis,"* 1–2.

reconsideration of the historical emergence of the hetero/homo binary and sexual identity."[86] It is significant that Kahan views Kaan's text as a way to further the study of the historical emergence of sexology in terms of sexual object choice and sexual identity. In the remainder of his introduction, Kahan returns to the question of the emergence of sexual perversions and identities. I would suggest, however, that Kaan's work can be read through the lens of sexual appetite.

Kaan intimated that he was compelled to write *Psychopathia Sexualis* due to the widespread sickness and the lack of available information to physicians and to the public:

> The great number of sick whom everywhere I saw corrupted by this disease ... the as yet smaller number of books that have been written on this disease—all these aroused in me a desire to collect case studies, to examine them and from them deduce general principles, and then to apply to them every kind of theoretical and practical knowledge and, thus, to derive from them rules useful to physicians.[87]

Kaan, however, included few case histories in his treatise. One of the more detailed cases appears towards the end of the book: the case of Mauritius S, an eighteen-year-old man who "became given over to onanism." In detailing the case, Kaan associated the patient's onanism with his visits to a gymnasium and his belonging to "the military institution." For Kahan, these notes constituted indications of the patient's homosexuality and suggested that the case of Mauritius S was one of the first scientific case studies of homosexuality. While I do not dispute this reading, the case also contained details that are reflective of an attention to the problem of amount in sexual appetites:

> His premature puberty and way of life certainly contributed much to developing the diseased seed. The attack of the disease itself sufficiently indicated its origin from this source, like a serpent afflicting all his organs and systems ... Since the evil was increasing daily, he was transferred from the homeopathic hospital into an allopathic one.[88]

[86] Ibid., 2.
[87] Ibid., 31–32.
[88] Ibid., 156.

The treatments provided to Mauritius S are reminiscent of humoral medicine—for example, the use of "light healing nutrients"[89]—a feature that is recurrent in Kaan's treatise. However, the condition of Mauritius S continued to worsen, his strength was "consumed"[90] and he passed away. The case notes, combined with Kaan's positioning of the imagination as both essential to sexuality yet prone to excessiveness through onanism, suggested that Mauritius S was more than a case of perversion. Rather, in the form of uncontrollable onanism, he overindulged his sexual appetites, which themselves were buttressed by uncontrollable imagination.

Sexual appetite was the force propelling individuals to abnormal acts. Importantly, Kaan's oeuvre reveals "the privileged link that exists between the sexual instinct and *phantasia* or imagination."[91] He wrote, for example, that "in every case of sexual aberration a morbidly aroused imagination holds sway, which clouds the mind. It seems neither absurd nor false [to] collect all such states as disease of the imagination."[92] The imagination is approached as a nearly autonomous force on which the subject has little control and the sexual sciences, from their inception, involved a particular disciplining of the imagination. Sexology, as Niklaus Largier argues, attempted to "eliminate the conspiratorial connection between imagination, affect, and libido."[93] Thus, sexual appetite in Kaan's thought was intimately connected to the imagination because of its natural propensity towards excessive sexual appetite. In fact, excess was at the centre of pathological sexuality itself. Kaan's classification of perversions relied heavily on notions of balance and degrees in the diagnosis and management of sexual appetite.

The association between a pathological imagination and excessive sexual appetite was made prominent through case studies published by nineteenth-century sexologists. However, it is worth highlighting that the connections between sexual lack and the imagination also appeared in Krafft-Ebing's cases. K, for instance, wanted marriage but "only on rational grounds" and was diagnosed with *anaesthesia sexualis*. Erotic novels made no impression on K and his familiarity with sexual life was limited to literature and stories heard from other men. He never masturbated and

[89] Ibid., 158.

[90] Ibid.

[91] Foucault, *Abnormal*, 280.

[92] Kahan, *Heinrich Kaan's "Psychopathia Sexualis,"* 82.

[93] Niklaus Largier, *In Praise of the Whip: A Cultural History of Arousal* (New York: Zone Books, 2007), 446.

had no interest in men or women. Krafft-Ebing reported no defects in K's psyche but mentioned the patient's fondness for solitude as well as his disinterest in the arts and in "the beautiful."[94] Krafft-Ebing himself used the trope of literature to describe *anaesthesia sexualis*: "a blank leaf in the life of an individual."[95] The metaphor of absence and lack evoke a missing or unwritten page in an individual life's story, and the physician's use of metaphor demonstrates the intimate connection between literature and sexology. "Since his seventeenth year he had at intervals nocturnal pollutions, but without concomitant lascivious dreams."[96] K's arid dreams revealed that his imagination was not functioning normally because no erotic fantasies were being produced. This disclosed the extent of K's pathological rationality.

Krafft-Ebing conceded that sexual appetite was subject to fluctuations, although he maintained that its presence was necessary. However, as he wrote earlier in *Psychopathia Sexualis*; "If man were deprived of sexual distinction and the nobler enjoyments arising therefrom, all poetry and probably all moral tendency would be eliminated from his life."[97] Krafft-Ebing did not need to enquire into K's fantasies; however, the imagination of the patient was still scrutinised. The patient needed to speak of his imagination and his dreams, as they were relevant to diagnosis. The genre of the case history sustained the promise of a life laid bare, a life revealed in its singularity and distinctiveness.[98] The fact that K did not exercise his imagination at all was pathological to the extent that even his dreams were desiccated. The imagination can also only be manifested in the patient case history. The imagination, if used over or below an optimum level, can lead the individual astray—too lacking or too excessive—showing how imagination itself is inextricable from pathologies of the sexual appetites. The arousing images must not "explode the framework of … naturalized sexuality"[99]; however, drawing on the case studies above, it can be argued

[94] Krafft-Ebing, *Psychopathia Sexualis*, 90–91.

[95] Ibid., 93.

[96] Krafft-Ebing, *Psychopathia Sexualis*, 91. The consideration of K's dreams foreshadows Sigmund Freud who would go on to place enormous significance on dreams and the psyche. On the influence of nineteenth-century sexologists on Freud, see Sulloway, *Freud, Biologist of the Mind*, 277–319.

[97] Krafft-Ebing, *Psychopathia Sexualis*, 29.

[98] John Forrester, "If *p* then what? Thinking in Cases," *History of Human Sciences* 9, no. 3 (1996): 10.

[99] Largier, *In Praise of the Whip*, 434.

that the imagination, images and "memory-pictures,"[100] as Schrenck-Notzing called them, need to exist in some way for a normal, balanced amount of sexual appetite.

The concept of narrative has several layers of meaning in sexological and psychiatric works of the nineteenth century. The patient first gave an account of their life, of the facts and key events that structured this life, and if it was disjointed, the psychiatrist would then make sense of this life and edit it in clinically intelligible language. The patient was also invited to speak of their fantasies, events that may not have occurred but are germane to assessing their sexual appetite. The clinician made connections between the imagined scenarios and the pathology of the patient. So, too, the clinician drew links between the *absence* of imagined plots and the pathology of the patient. I argue, then, that an ideal amount of sexual appetite was envisioned through this process of constructing a case history, at the nexus of speech, imagination and writing.

The case of Z discussed in previous pages brought together Krafft-Ebing's concern with an excessive imagination and homosexuality. Z intended to marry but found himself attracted to "male performers" when visiting circuses and theatres. He frequently indulged in "excessive masturbation," which made him neurasthenic. Krafft-Ebing noted the following: "frequent pollutions in sleep occurred … It was only occasionally that he dreamed of men when he had pollutions; and never of women … He dreamed of death-scenes, of being attacked by dogs, etc. After these, as before, he suffered with great *libido sexualis*."[101] The case history of Z demonstrated an attention to dreams, imagination, masturbation, homosexuality and sexual excess. In *Therapeutic Suggestions*, Schrenck-Notzing, who devoted a large part of his work to the study of dreams, also often approached masturbation as a problem of excess. "Another patient, who is at the time of contrary sexuality, has erection [sic.] only when he sees naked male nates. The idea of the nates is not sufficient, at least, in the waking state, though it is efficient in dreams."[102] The medicalisation of sexual appetite and its connections to an excessive imagination clearly became most effectively narrated through the patient case history.

[100] Schrenck-Notzing, *Therapeutic Suggestions*, 69.

[101] Krafft-Ebing, *Psychopathia Sexualis*, 376.

[102] Schrenck-Notzing, *Therapeutic Suggestions*, 13. "Nates" is an archaic term for buttocks.

This chapter has shown that the intermingling of the imagination, appetite and sexuality has a complicated history. In eighteenth-century anti-onanism literature, the imagination was positioned as especially problematic. Exercising one's imagination could lead to a range of excesses, which usually began with the act of self-pleasure. In turn, excessive amounts of sexual appetite would quickly become uncontrollable and eventually result in sexual lack. In addition, as Laqueur reminds us, such concerns were often connected to sexual object choice.[103] The attention to the fluctuations of appetite and their connections to sexual perversions, while circulating within different theories and concepts in a range of physicians' works, coalesced to condition the emergence of the *scientia sexualis*.

Throughout the nineteenth century, information on patients' inner lives was harnessed in different ways: teratological examinations, designing family trees of inherited diseases, anamneses, editing patients' words and combining them with theories of sexuality. Excessive amounts of sexual appetite were inextricable from problematisations of the imagination in the case history. The medicalisation of sexual appetite emerged as a problem of the imagination; indeed, they were often inseparable, and their problematisation was mediated through the technique of the case history. In the second half of the nineteenth century, sexuality was no longer exclusively restricted "to the anatomical structure of the internal and external genital organs," as Davidson argues. "It is now a matter of impulses, tastes, aptitudes, satisfactions, and psychic traits."[104] While the bulk of scholarly research on the genealogy of sexuality has situated the nineteenth century as a time when sexual object choice becomes pathologised, this chapter has examined how sexual appetite acted as a structuring instrument to sexual perversions. The dynamics of appetite provided a network of intensity, duration and moderation that served to medicalise disorders such as sexual inversion.

This chapter has examined how the medicalisation of sexual appetite in nineteenth-century sexology emerged through the technique of the patient case history. This technique presented sexual appetite as a structuring device in the expansion of taxonomies of sexual perversions and became a tool to uncover the depths of perverse imaginations. In the next chapter, the book explores how techniques of elixirs and devices emerged alongside the patient case history in the management of sexual appetite in

[103] Laqueur, *Solitary Sex*, 264.
[104] Davidson, *The Emergence of Sexuality*, 35.

the nineteenth century. The United States in the late-nineteenth and early-twentieth centuries was a fertile era for the production of new techniques and discourses of sexual appetite, in particular for "healing" so-called lost manhood. Chapter 3 thus explores how products claiming to provide solutions for reviving "manly vigor" were advertised in newspapers and marriage manuals, and the way in which prospective consumers informed themselves on how to manage their sexual lives. The subject who emerged at the beginning of the twentieth century was not simply a patient, but a customer who needed both a product and knowledge to manage and control their sexuality.

Bibliography

Bauer, Heike. 2009. *English Literary Sexology: Translations of Inversion 1860–1930.* Basingstoke: Palgrave Macmillan.

———. 2015. *Sexology and Translation: Cultural and Scientific Encounters Across the Modern World.* Philadelphia: Temple University Press.

Berkenkotter, Carol. 2008. *Patient Tales: Case Histories and the Uses of Narrative in Psychiatry.* Columbia, SC: University of Southern Carolina Press.

Binet, Alfred. 1888. *Le fétichisme dans l'amour.* Paris: Octave DOIN.

Bloch, Iwan. 1909 [1906]. *The Sexual Life of Our Time in Its Relations to Modern Civilization.* Trans. M Eden Paul. London: Rebman Limited.

Bristow, Joseph. 1998. Symonds's History, Ellis's Heredity: *Sexual Inversion.* In *Sexology in Culture: Labelling Bodies and Desires*, eds. Lucy Bland and Laura Doan, 79–99. Cambridge: Polity Press.

Crozier, Ivan. 2000. Havelock Ellis, Eonism and the Patient's Discourse; or, Writing a Book About Sex. *History of Psychiatry* 11 (42): 125–154.

———, ed. 2008. *Havelock Ellis and John Addington Symonds, Sexual Inversion: A Critical Edition.* Basingstoke: Palgrave Macmillan.

———. 2008. Pillow Talk: Credibility, Trust and the Sexological Case History. *History of Science* 46 (154): 375–404.

Cryle, Peter, and Alison Moore. 2011. *Frigidity: An Intellectual History.* London: Palgrave Macmillan.

Cryle, Peter, and Elizabeth Stephens. 2017. *Normality: A Critical Genealogy.* Chicago: University of Chicago Press.

Davidson, Arnold I. 2001. *The Emergence of Sexuality: Historical Epistemology and the Formation of Concepts.* Cambridge, MA: Harvard University Press.

Didi-Huberman, Georges. 2003. *Invention of Hysteria: Charcot and the Photographic Iconography of the Salpêtrière.* Trans. Alisa Hartz. Cambridge, MA: MIT Press.

Dowbiggin, Ian R. 1991. *Inheriting Madness: Professionalization and Psychiatric Knowledge in Nineteenth-Century France*. Berkeley: University of California Press.
Ellis, Havelock. 1942 [1905]. *Studies in the Psychology of Sex, Volume 1*. London: William Heinemann Medical Books.
Epstein, Julia. 1992. Historiography, Diagnosis, and Poetics. *Literature and Medicine* 11 (1): 23–44.
Forrester, John. 1996. If *p* then what? Thinking in Cases. *History of Human Sciences* 9 (3): 1–25.
Foucault, Michel. 1978. *The Will to Knowledge: The History of Sexuality, Volume 1*. Trans. Robert Hurley. London: Penguin Books.
———. 1991 [1977]. *Discipline and Punish: The Birth of the Prison*. Trans. Alan Sheridan. Middlesex: Penguin Books.
———. 2003. *Abnormal: Lectures of the Collège de France, 1974–1975*. Trans. Graham Burchell. New York: Picador.
Freud, Sigmund. 2001 [1905]. *The Standard Edition of the Complete Psychological Works of Sigmund Freud, Volume VII*. Trans. James Stratchey. London: Vintage Books.
Germon, Jennifer. 2009. *Gender: A Genealogy of an Idea*. New York: Palgrave Macmillan.
Gillis, Jonathan. 2006. The History of the Patient History Since 1850. *Bulletin of the History of Medicine* 80 (3): 490–512.
Goldstein, Jan. 1981. *Console and Classify: The French Psychiatric Profession in the Nineteenth Century*. Cambridge: Cambridge University Press.
Hekma, Gert. 1991. A History of Sexology: Social and Historical Aspects of Sexuality. In *From Sappho to de Sade: Moments in the History of Sexuality*, ed. Jan Bremmer, 173–193. New York: Routledge.
———. 1993. 'A Female Soul in a Male Body': Sexual Inversion as Gender Inversion in Nineteenth-Century Sexology. In *Third Sex, Third Gender: Beyond Sexual Dimorphism in Culture and History*, ed. Gilbert Herdt, 213–239. New York: Zone Books.
Herzer, Manfred. 1986. Kertbeny and the Nameless Love. *Journal of Homosexuality* 12 (1): 1–26.
Hunter, Katherine Montgomery. 1991. *Doctor's Stories: The Narrative Structure of Medical Knowledge*. Princeton: Princeton University Press.
Hutter, Jörg. 1993. The Social Constructions of Homosexuals in the Nineteenth Century: The Shift from the Sin to the Influence of Medicine in Criminalizing Sodomy in Germany. *Journal of Homosexuality* 24 (3–4): 73–93.
Kahan, Benjamin, ed. 2016. *Heinrich Kaan's "Psychopathia Sexualis" (1844): A Classic Text in the History of Sexuality*. Trans. Melissa Haynes. Ithaca: Cornell University Press.
Katz, Jonathan Ned. 1995. *The Invention of Heterosexuality*. New York: Dutton.

Kennedy, Hubert. 1988. *Ulrichs: The Life and Works of Karl Heinrich Ulrichs, Pioneer of the Modern Gay Movement.* Boston: Alyson Publications.

Krafft-Ebing, Richard von. 1965 [1886]. *Psychopathia Sexualis: A Medico-Forensic Study.* Trans. Harry E Wedeck. New York: G. P. Putnam's Sons.

Laqueur, Thomas. 1989. Bodies, Details and the Humanitarian Narrative. In *The New Cultural History*, ed. Lynn Hunt, 176–204. Berkeley: University of California Press.

———. 2003. *Solitary Sex: A Cultural History of Masturbation.* New York: Zone Books.

Largier, Niklaus. 2007. *In Praise of the Whip: A Cultural History of Arousal.* New York: Zone Books.

Magnan, Valentin. 1885. *Des anomalies, des aberrations et des perversions sexuelles.* Paris: A. Delahaye & E. Lecrosnier.

———. 1893. *Recherches sur les centres nerveux: Alcoolisme, folie des héréditaires dégénérés, paralysie générale, médecine légale.* Paris: G. Masson.

Matysik, Tracie. 2004. In the Name of the Law: The "Female Homosexual" and the Criminal Code in Fin de Siecle Germany. *Journal of the History of Sexuality* 13 (1): 26–48.

McWhorter, Ladelle. 1999. *Bodies and Pleasures: Foucault and the Politics of Normalization.* Bloomington: Indiana University Press.

Moll, Albert. 1893. *Les perversions de l'instinct génital: Étude sur l'inversion sexuelle.* Trans. Dr Pactet. Paris: Georges Carré.

———. 1912 [1909]. *The Sexual Life of the Child.* Trans. Eden Paul. New York: The Macmillan Company.

Moore, Alison. 2009. The Invention of Sadism? The Limits of Neologisms in the History of Sexuality. *Sexualities* 12 (4): 486–502.

Morel, Bénédict Augustin. 1857. *Traité des dégénérescences physiques, intellectuelles et morales de l'espèce humaine et des causes qui produisent ces variétés maladives.* Paris: J. B. Baillière.

Mosse, George L. 1983. Nationalism and Sexuality in Nineteenth-Century Europe. *Culture & Society* 20: 75–84.

Nordau, Max Simon. 1895 [1892]. *Degeneration.* New York: D Appleton & Company.

Nowell-Smith, Harriet. 1995. Nineteenth-Century Narrative Case Histories: An Inquiry into the Stylistics and History. *Canadian Bulletin of Medical History* 12: 47–67.

Nye, Robert A. 1991. The History of Sexuality in Context: National Sexological Traditions. *Science in Context* 4 (2): 387–406.

———. 1993. The Medical Origins of Fetishism. In *Fetishism as Cultural Discourse*, eds. Emily Apter and William Pietz, 13–30. Ithaca and London: Cornell University Press.

Oosterhuis, Harry. 2000. *Stepchildren of Nature: Krafft-Ebing, Psychiatry and the Making of Sexual Identity*. Chicago: University of Chicago Press.

———. 2019. Albert Moll's Ambivalence About Homosexuality and His Marginalization as a Sexual Pioneer. *Journal of the History of Sexuality* 28 (1): 1–43.

Pick, Daniel. 1989. *Faces of Degeneration: A European Disorder, c.1848–1918*. Cambridge: Cambridge University Press.

Reed, Matt. 2004. *La manie d'écrire*: Psychology, Auto-Observation and Case History. *Journal of the History of Behavioral Sciences* 40 (3): 265–284.

Schaffner, Anna Katharina. 2012. *Modernism and Perversion: Sexual Deviance in Sexology and Literature, 1850–1930*. Basingstoke: Palgrave Macmillan.

Schrenck-Notzing, Albert von. 1895. *Therapeutic Suggestion in Psychopathia Sexualis, with Especial Reference to Contrary Sexual Instinct*. Trans. Charles Gilbert Chaddock. Philadelphia: The F. A. Davis Company.

Sealey, Anne. 2011. The Strange Case of the Freudian Case History: The Role of Long Case Histories in the Development of Psychoanalysis. *History of Human Sciences* 24 (1): 36–50.

Sulloway, Frank J. 1979. *Freud, Biologist of the Mind: Beyond the Psychoanalytic Legend*. New York: Basic Books.

Takács, Judit. 2004. The Double Life of Kertbeny. In *Past and Present of Radical Sexuality Politics*, ed. Gert Hekma, 26–40. Amsterdam: Mosse Foundation.

Ulrichs, Karl Heinrich. 1994. *Riddle of Man-Manly Love: The Pioneering Work on Male Homosexuality*. Trans. Michael A Lombardi-Nash. Buffalo: Prometheus Books.

Warner, John Harley. 1999. The Uses of Patient Records by Historians: Patterns, Possibilities and Perplexities. *Health and History* 1 (2–3): 101–111.

Waters, Chris. 1998. Havelock Ellis, Sigmund Freud and the State: Discourses of Homosexual Identity in Interwar Britain. In *Sexology in Culture: Labelling Bodies and Desires*, eds. Lucy Bland and Laura Doan, 165–179. Cambridge: Polity Press.

Weeks, Jeffrey. 2012 [1981]. *Sex, Politics and Society: The Regulation of Sexuality Since 1800*. London: Routledge.

Wolffram, Heather. 2009. *The Stepchildren of Science: Psychical Research and Parapsychology in Germany, c. 1870–1939*. Amsterdam: Rodopi B.V.

CHAPTER 3

Elixirs of Vigour

The late-nineteenth century witnessed the emergence of a modern advertising industry in Western societies. The rapid expansion of the economies of the United States, France and England during the industrial revolution democratised the dissemination and accessibility of books, manuscripts, periodicals and newspapers.[1] In the United States, the combination of cheap paper, rapid printing techniques and the construction of mass transport networks popularised the newspaper and the periodical as authoritative sources of advertisements. Nearly all occupations, leisure and interest groups produced their own weekly or monthly publications featuring adverts for different products, services and equipment.[2] It was also common during this era to observe advertisements for often-dubious medical products purporting to cure all kinds of afflictions.[3] In late Victorian America, sexual appetite, especially male sexual potency, appeared as a recurring aspect of the advertising industry. Qualified physicians, laypeople

[1] See Elizabeth Eisenstein, *The Printing Press as an Agent of Change* (Cambridge: Cambridge University Press, 1979), Eltjo Buringh and Jan Luiten van Zanden, "Charting the 'Rise of the West': Manuscript and Printed Books in Europe, a Long-Term Perspective from the Sixth through Eighteenth Centuries," *The Journal of Economic History* 69, no. 2 (2009): 409–445, and William G. Gabler, "The Evolution of American Advertising in the Nineteenth Century," *The Journal of Popular Culture* XI, no. 4 (1978): 763–771.

[2] Gabler, "Evolution of American Advertising," 767.

[3] Jane Marcellus, "Nervous Women and Noble Savages: The Romanticized 'Other' in Nineteenth Century US Patent Medicine Advertising," *The Journal of Popular Culture* 41, no. 5 (2008): 787.

J. Flore, *A Genealogy of Appetite in the Sexual Sciences*,
https://doi.org/10.1007/978-3-030-39423-3_3

and quacks[4] supplied products to an eager population, and commercials for "manly vigor," "lost vigor" or "lost manhood" appeared frequently in daily papers. It is during the long nineteenth century that the marketing and commercialisation of low sexual appetite in men—commonly referred to as "impotence"—began to prosper.[5]

Merchandise was not limited to newspapers; advertising featured on the pages of marriage guides and sex manuals authored by moral educators, physicians and quacks alongside various remedies for "lost manhood." The products offered by writers varied from concoctions and potions, to ointments and mechanical devices. The emphasis in advice literature on marriage and sexuality was on providing tools for self-control.[6] Low and excessive sexual appetite were regularly problematised in the marital sex manual.[7] Although the focus fell quite sharply on men, women's sexual vigour was often discussed in those publications as needing management. Through a concern with the governance of sexual appetite, newspapers, manuals and tracts advertised elixirs that promised to stir or allay the passions. As such, the various products marketed presented potential consumers with the possibility to take control of their sexual appetite.

This chapter traces the circulation of products in the medicalisation of sexual appetite in the late-nineteenth and early-twentieth centuries. It examines the commodification of sexual appetite through the advertisements of aphrodisiacs, elixirs and mechanical devices in manuals, pamphlets, tracts and newspapers. The subject who emerged at the beginning of the twentieth century was not simply a patient, but a customer who needed both a product and knowledge to manage and control their sexuality. Patients were no longer considered passive subjects of medical

[4] See Arthur Wrobel, ed., *Pseudo-Science and Society in 19th-Century America* (Lexington: The University Press of Kentucky, 2015).

[5] Kathleen L. Endres, "From 'Lost Manhood' to 'Erectile Dysfunction': The Commercialization of Impotence" in *We Are What We Sell: How Advertising Shapes American Life…and Always Has, Volume 2: Advertising at the Center of Popular Culture: 1930s–1975*, eds. Danielle Sarver Coombs and Bob Batchelor (Santa Barbara, CA: Praeger, 2014), 85.

[6] M. E. Melody and Linda M. Peterson, *Teaching America about Sex: Marriage Guides and Sex Manuals from the Late Victorians to Dr. Ruth* (New York: New York University Press, 1999), 21.

[7] I borrow the term "marital sex manual" from Jessamyn Neuhaus as it encompasses the conjugal couple and the proper management of sexuality more broadly. Jessamyn Neuhaus, "The Importance of Being Orgasmic: Sexuality, Gender, and Marital Sex Manuals in the United States, 1920–1963," *Journal of the History of Sexuality* 9, no. 4 (2000): 447–473.

diagnoses and instead became active consumers utilising a range of techniques to both enhance and govern sexual appetite. To this extent, the chapter considers how the marketing of aphrodisiacs, elixirs and mechanical devices for sexual imbalances were used for managing sexual appetite. The different products promoted in this era functioned as techniques for actively encouraging individuals to autonomously and independently manage their sexual lives. Whether individuals were encouraged to consume foods, herbal remedies or instruments, the narrative of responsibility and self-improvement permeated the marketing and use of those products.

Nourishment and the Dietetics of Married Life

Alongside the growth of the newspaper industry, the late-nineteenth century witnessed the emergence of marital sex advice literature. This popular field of literature drew from theology, natural philosophy and medicine to counsel individuals on the management of sexual appetite.[8] From tracts admonishing sexual excesses to small pamphlets emphasising pleasure, technique, and the mutual orgasm, marriage manuals constituted repositories of knowledge on erotic and sexual customs. They functioned as tools for individuals to achieve mastery over their bodies and intimate relations, and for the conjugal couple to manage domestic life and attain balance in the performance of sexual activity. Whether writers were discussing the dangers that accompanied carnal acts in the nineteenth century, or the benefits of simultaneous orgasms in the twentieth century, the necessity to manage sexual appetite and its flows was a persistent issue. The manuals—encompassing advice on sex and other aspects of the domestic sphere—provided insights into sexual practices during these centuries, or at least they provided a window into how so-called experts believed individuals should behave.[9] Similarly, Michael Gordon observes that marriage manuals "provide us with a picture of what 'experts' felt to be desirable patterns at a particular time."[10] It is indeed difficult to determine what kind of influence those works had on individuals. However,

[8] See Michael Gordon and M. Charles Bernstein, "Mate Choice and Domestic Life in the Nineteenth-Century Marriage Manual," *Journal of Marriage and Family* 32, no. 4 (1970): 666–667.

[9] Ronald G. Walters, *Primers for Prudery: Sexual Advice to Victorian America* (Baltimore, MD: Johns Hopkins University Press, 2000), 11.

[10] Michael Gordon, "The Ideal Husband as Depicted in the Nineteenth Century Marriage Manual," *The Family Coordinator* 18, no. 3 (1969): 226.

even if the manuals "do not necessarily reflect private behaviour," they certainly "can place sex in its network of cultural norms."[11]

The problematisation of the body was a central feature of advice literature. Writers of manuals, pamphlets and advertisements urged individuals to manage and care for their bodies, because a common concern at the time was the fear of loss of bodily autonomy and integrity through male disorders such as spermatorrhoea. French physician Claude-François Lallemand identified spermatorrhoea by drawing on two popular paradigms of the era: physical debility and sperm as an essential yet limited fluid.[12] Defined as an excessive and uncontrollable discharge of sperm, spermatorrhoea was believed to cause nervousness, impotence, fatigue and even death. Lallemand attributed the causes of the disease to masturbation and excessive sexual activity. It was a disorder of excess *par excellence*, as it evoked the possibility of a complete loss of control, a body in a perpetual state of excess.

The literature at the time demonstrated a marked concern with the potency and vigour of men. While women certainly featured prominently in medical and popular writings on sex, their concerns were frequently superseded by a problematisation of low sexual appetite in men. Historically, low sexual appetite, or impotence, was variously conceived as a form of divine retribution for sin, as barrenness, and a consequence of excessive sexual activity.[13] From the 1750s to the 1850s, Kevin J. Mumford notes, "the dominant conception of impotence shifted from predominantly a problem of fertility to a problem of diminished sexual capacity."[14] Impotence in men became a physical debility, rather than a religious

[11] Walters, *Primers for Prudery*, 11.

[12] Claude-François Lallemand, *A Practical Treatise on the Causes, Symptoms and Treatment of Spermatorrhoea*, trans. Henry J. McDougall (Philadelphia: Blanchard and Lea, 1861 [1840]). See also Ellen Bayuk Rosenman, "Body Doubles: The Spermatorrhea Panic," *Journal of the History of Sexuality* 12, no. 3 (2003): 365–399, and Elizabeth Stephens, "Pathologizing Leaky Male Bodies: Spermatorrhea in Nineteenth-Century British Medicine and Popular Anatomical Museums," *Journal of the History of Sexuality* 17, no. 3 (2008): 421–438.

[13] See Peter Cryle and Alison Moore, *Frigidity: An Intellectual History* (London: Palgrave Macmillan, 2011), and Angus McLaren, *Impotence: A Cultural History* (Chicago and London: University of Chicago Press, 2007).

[14] Kevin J. Mumford, "'Lost Manhood' Found: Male Sexual Impotence and Victorian Culture in the United States," *Journal of the History of Sexuality* 3, no. 1 (1992): 37.

issue.[15] It should be noted that while the paradigm of physical disorder undoubtedly became prominent at this time, it was widely associated with external factors such as entertainment, literature and urban life. Debility became key to the conceptualisation of impotence as writers, social reformers and physicians decried the demands of civilisation and the spreading of "nervous disorders."[16]

Notions of men as lustful and women as both passive and guardians of societal morality were prominent in the nineteenth century. As Mumford highlights, male sexual impotence was intimately connected to ideas of "overcivilisation." Middle-class men were particularly vulnerable to impotence. While male sexual appetite was often discussed in terms of fertility, it is also their masculinity that was at stake.[17] For Mumford, an analysis of the historical transformations from lost manhood to impotence "not only illuminates the direction and forces of sexual change, but it also suggests [that] definitions of male sexuality were shaped by, and in turn reinforced, systems of inequality."[18] Hence, my examination of lost manhood in this chapter focuses on how ideas of quantity, excess and lack underpinned conceptualisations of sexuality itself, and how elixirs and devices were harnessed as techniques of knowledge in the formation of sexual subjectivities.

Manuals instructed men on avoiding the depletion of their sexual appetite and managing their fertility. Men and women were both believed to suffer from the consequences of sexual excess, but men typically bore "the brunt of the dangers."[19] This was connected to the belief

[15] This paradigm would shift to repressed instincts in the early-twentieth century with the rise of psychoanalysis. Note that competing medical theories of the era also harnessed physical constitution. Phrenology, for example, discussed the perversions of "amativeness," where a protruding skull would reveal a stronger need for amorous activities, but a smaller "organ" would make "the person less susceptible to the emotions of love." L. N. Fowler, *Marriage: Its History and Ceremonies; with a Phrenological and Physiological Exposition of the Functions and Qualifications for Happy Marriages* (New York: Fowler & Wells, 1848), 78.

[16] See the works of the American physician and neurologist George Miller Beard, who extensively wrote on neurasthenia, the disorder of enfeebled nerves: *American Nervousness: Its Causes and Consequences, a Supplement to Nervous Exhaustion (Neurasthenia)* (New York: G. P. Putnam's Sons, 1881) and *Sexual Neurasthenia: Its Hygiene, Causes, Symptoms, and Treatment with a Chapter on Diet for the Nervous* (New York: E. B. Treat, 1884).

[17] Mumford, "'Lost Manhood' Found," 35.

[18] Ibid. 57.

[19] Michael Gordon, "From an Unfortunate Necessity to a Cult of Mutual Orgasm: Sex in American Marital Education Literature 1830–1940," in *Studies in the Sociology of Sex*, ed. James M. Henslin (New York: Meredith Corporation, 1971), 56.

that semen was a vital, precious and limited fluid that should not be wasted.[20] Advice literature would also include guidance on a "dietary" regime of sexual activity. Writers provided advice on the frequency of copulation based on a host of factors in the individual's life. For example, in his *Sexual Physiology and Hygiene* (1891), the American medical reformer R.T. Trall wrote in a section titled "Frequency of Sexual Intercourse":

> The frequency with which sexual intercourse can be indulged, without serious damage to one or both parties, depends, of course, on a variety of circumstances—constitutional stamina, temperament, occupation, habits of exercise, period of life, etc. Few should exceed the limit of once a week; while many cannot safely indulge oftener than once a month. But … temperance is always a safer rule of conduct.[21]

Writers also advocated continence, both to preserve essential fluids and as a form of contraception. Men were urged to exercise restraint from sexual activity, "holding up impotence as one extreme consequence of male sexual misconduct."[22] As exemplified by the quote from Trall above, the frequency of sexual activity was connected to occupation and habits of everyday life. Sexual appetite was rarely treated in isolation, as writers would counsel individuals to pay attention to other parts of their body and daily routines.

In the nineteenth century, literature purporting to discuss marriage advice focused strongly on health, hygiene and physiology. The major topics covered at the time included the perils of masturbation, conception, pregnancy and the function of genital organs.[23] In his cultural history of impotence, McLaren observes that impotence in men was often associated with onanism and excessive intercourse. The notion that sexual lack was a result of excessive behaviour was common and McLaren interprets this as reflecting physicians' (and quacks') concern for "youthful indiscretions, in particular masturbation."[24] I would add that this connection also reflects

[20] Ibid.

[21] R.T. Trall, *Sexual Physiology and Hygiene: An Exposition Practical, Scientific, Moral, and Popular, of Some of the Fundamental Problems in Sociology* (New York: M. L. Holbrook & Co., 1891), 233.

[22] Mumford, "'Lost Manhood' Found," 40.

[23] Gordon, "From an Unfortunate Necessity to a Cult of Mutual Orgasm," 55.

[24] McLaren, *Impotence*, 133.

the place that sexual appetite—in particular, frequency—had in ideas on sexuality and married life. Sexual appetite here refers not only to excess and lack as discrete categories but rather to their interrelation alongside concepts of balance, amount and frequency. Doctors and quacks were not solely concerned with offering products for "lost manhood," their discussions reveal a preoccupation with questions of frequency and balance in sexual activity. This, as I go on to discuss in this chapter, operated in tandem with techniques of self-governance. The risks posed by extramarital sex, and by extension sexual excess, were not solely popularised to consolidate the institution of marriage; such concerns were also reflective of an understanding of sexuality based on fluctuations, amounts and frequency.

The amount of sexual appetite and so-called over-indulgence were recurrent issues. In medical and pseudo-medical texts of the time, writers continually warned of the dangers of excessiveness while providing methods to manage appetite. The authors Jefferis and Nichols, for example, noted that while they did not wish to devise rules for married couple, they believed that, for the "best … government of the marriage-bed,"

> sexual indulgence should only occur about once in a week or ten days … it is a hygienic and physiological fact that those who indulge only once a month receive a far greater degree of the intensity of enjoyment than those who indulge their passions more frequently. Much pleasure is lost by excesses where much might be gained by temperance, giving rest to the organs for the accumulation or nervous force.[25]

The passage above, from *Search Lights on Health* (1894), reflects a concern with the depletion of sexual energy, a common idea at the time, as scholars historicising disorders such as spermatorrhoea and seminal weakness have remarked.[26] We find here the idea of a finite amount of "nervous force" that can potentially be squandered. However, frequency and amount also became problematised, as they emerged as aspects of sexuality needing good "government." Writers were suspicious of excessive sexual activity and moderation was regularly advised. In providing advice to readers on how to conduct oneself and attend to one's body, authors developed regimens of sex. The techniques and guidelines of self-management

[25] B.G. Jefferis and J.L. Nichols, *Search Lights on Health: Light on Dark Corners, A Guide to Purity and Physical Manhood. Advice to Maiden, Wife and Mother. Love, Courtship and Marriage* (Canada: The J L Nichols Company, 1894), 211.

[26] McLaren, *Impotence*, 134.

detailed in the manuals were attached to, and reflective of, dominant ideas on sex and gender. Advice literature on health, marriage and sexuality reveal prevalent ideas on the gendering of sexual inadequacy. These works demonstrate the societal expectations placed on men and women with regards to the uses of sexuality.

In her analysis of Victorian sexual customs, Nancy F. Cott uses the term "passionlessness" to explain the discursive productions of knowledge on female sexuality during this period. As Cott writes, women in the first half of the nineteenth century were generally thought to possess little sexual assertiveness, "their sexual appetites contributed a very minor part … to their motivations."[27] Ideas on female sexuality were not monolithic as notions of "fallen women" and "hysterical women" with rabid sexual appetites also featured in medical treatises and moral literature.[28] Cott suggests that the Western notion that women were exceptionally libidinous, popular prior to the nineteenth century, transformed into the view that women "were *less* carnal and lustful than men."[29] In addition, throughout the nineteenth century, female sexuality was widely discussed in terms of reproductive capacity. Thus, while medical knowledge on female sexuality was often contradictory, discussions on low sexual appetite in women were often accompanied by its consequences on sexual reproduction. Cott examines how women's supposed lack of "carnal motivation" influenced ideas of their moral superiority and was employed to widen their political and social opportunities.[30] Carolyn J. Dean also comments that Victorian ideas on female sexuality were contradictory. Women were "sexually passive and hypersexual."[31] Although their passivity represented their potential for achieving a "higher good," it was nonetheless taken as a lack of self-control as it made them more docile.[32]

The mechanisms or tools provided to individuals in manuals, but also the manuals themselves, produced, or at least aimed to produce, governable subjects. The subject of marital sex advice literature was provided

[27] Nancy F. Cott, "Passionlessness: An Interpretation of Victorian Sexual Ideology," *Signs: Journal of Women in Culture and Society* 4, no. 2 (1978): 220.

[28] See Carroll Smith-Rosenberg, *Disorderly Conduct: Visions of Gender in Victorian America* (New York: Oxford University Press, 1985), 197–216.

[29] Cott, "Passionlessness," 221 (emphasis original).

[30] Ibid., 233.

[31] Carolyn J. Dean, *Sexuality and Modern Western Culture* (New York: Twayne Publishers, 1996), 6.

[32] Ibid.

with the necessary knowledge to conduct her/his life appropriately. While the guidelines were sometimes quite prescriptive, a responsibilisation of the subject also circulated where the individual was encouraged to consume certain products and apply this knowledge to everyday life. Written documentation encompassing advice literature, newspaper advertising and medical texts, emphasised individual responsibility and social health. Individuals had the responsibility to exercise self-restraint. Using the language of morality and temperance,[33] writers urged their audience to practise self-control for their own good and for the welfare of society. In other words, self-restraint was framed as part of the social contract. Discussions on sexuality and health appeared alongside guidelines on the administration of the family and the broader organisation of society. Topics such as sleeping arrangements, domestic duties and diet were also included. At a time where restraint was a guiding principle of social life, sexuality and diet were closely intertwined. This led the American social reformer Sylvester Graham to claim that "[d]igestion was the first physiological process to suffer the effects of sexual excess."[34]

Throughout the early part of the nineteenth century, the purpose of sexual activity remained intimately connected to generation. Consequently, the effect of certain foods and drinks was often discussed in terms of fertility. This is not to suggest that sexual vigour was absent from this discourse, but that authors often approached vigour and energy as an issue of fertility. This perspective was reflected in different parts of the Anglo-Saxon world. For instance, the British physician Michael Ryan wrote in his *Philosophy of Marriage* (1837), "Experience has also shown that certain foods excite the genital organs of particular individuals. The employment of vinous and spirituous liquors produces the same effect on most persons, but their abuse, as also that of warm drinks, such as tea and coffee, are injurious to generation."[35]

[33] On the temperance movement and sexuality in the United States, John D'Emilio and Estelle B. Freedman, *Intimate Matters: A History of Sexuality in America*, 3rd ed (New York: Harper & Row, 2012 [1988]).

[34] Stephen Nissenbaum, *Sex, Diet, and Debility in Jacksonian America: Sylvester Graham and Health Reform* (Westport, CT: Greenwood Press, 1980), 107.

[35] Michael Ryan, *The Philosophy of Marriage, in its Social, Moral, and Physical Relations* (London: John Churchill, 1837), 149.

The stability of the home and the family depended on ability and potency[36] and infertility represented a threat to this central institution and by extension to society. Indeed, the management of frequency and amounts became part of the power relations involved in sexuality. Foods and drinks deemed to possess aphrodisiac powers were discussed as means to enhance and promote generation. As Jennifer Evans comments, "for many early modern men and women, using [aphrodisiacs] to provoke sexual desire was considered a way of improving fertility."[37] It is worth noting that Evans' study of aphrodisiacs in early-modern England focuses on how sexual stimulants were used to promote fertility and treat barrenness. While she considers how aphrodisiacs were consumed to regulate sexual desire, her focus on procreation turns the analysis to sexual object choice and coupled heterosexuality, rather than a consideration of how elements of lack and excess connected to sexuality itself and what techniques were mobilised to encourage effective self-governance.

As this book examines, from early on, nourishment played an important part in the management of sexual appetite. Writers were not only concerned with external influences, such as climate and temperature; they also considered important what individuals put into their bodies. The techniques of the management of sexual appetite thus involved a problematisation of bodily functions. The authors of marital sex advice texts reflected on the internal processes of the body, how the body made use of food and what sorts of internal effects food had on sexuality and, by extension, on respectable conduct. The sensory experience of food was frequently problematised and became connected to embodied experience and social health.

To prevent the loss of sexual vigour, authors of manuals would, among other solutions, counsel a change in dietary habits. George Miller Beard, the American neurologist best known for his treatises on neurasthenia, devoted part of *Sexual Neurasthenia* (1884) on the "diet of the nervous." Regulating nerves, and sexuality more broadly, inevitably necessitated a monitoring of what the individual ingests. He claimed, "[f]ood is medicine"[38]—a statement that resonated in medical and popular discourse on sexuality. The treatment of nerves and the regulation of sexual activity,

[36] Jennifer Evans, *Aphrodisiacs, Fertility and Medicine in Early Modern England* (Suffolk: The Boydell Press, 2014), 25.

[37] Ibid., 11.

[38] Beard, *Sexual Neurasthenia*, 248.

for Beard, were only effective if accompanied by a proper diet, otherwise the sufferer would be condemned to nervous debilities. Neurasthenia was discussed in connection to both excessive and low sexual appetite. Beard in fact treated diminished sexual appetite as both a matter of degree and an inability to exert control over one's body (e.g. through the disorder of priapism). In this same work, Beard also counselled that it might be necessary for those afflicted to "sleep apart."[39]

Achieving continence and moderation was connected to nourishment; what beverages and foods individuals ingested would contribute to a stronger governance of sexuality. The breakfast inventor, John Harvey Kellogg claimed in *Plain Facts about Sexual Life* (1877) that condiments had overly stimulating effects: "In addition to the indirect injury which is done to the sexual organs by condiments through disturbance of stomach and liver, very many condiments have a direct influence in occasioning excitement and congestion of those parts."[40]

Plain Facts was intended as a familial educational manual. For Kellogg, stimulating foods affected sexual health and sexual appetite. The selection of food and beverages was an essential aspect of self-control and was central to leading a temperate life. Kellogg created "Corn Flakes" as a healthy breakfast option and, as Rosalyn M. Meadow and Lillie Weiss suggest, "an extinguisher of sexual desire, since it was then believed that wheat was a stimulant with aphrodisiac properties."[41] Some authors, such as Trall, went as far as definitively attributing sexual excess to improper nutrition: "[the] enormous and unnatural development of the sexual passions are largely the effect of highly-stimulating foods and drinks. Alcohol and tobacco no doubt goad this instinct into such a fever that it is almost uncontrollable."[42]

[39] See "Marrying and Not Marrying" in *Sexual Neurasthenia*, 130–132.

[40] John Harvey Kellogg, *Plain Facts about Sexual Life* (Battle Creek, MI: Office of the Health Reformer, 1877), 340. Alice B. Stockham issued similar advice to women: "To live continent lives, avoid food containing aphrodisiac stimulants, such as coffee, eggs, oysters, and animal food. *Omit the evening meal*; for the purpose desired this stands paramount to all other means. Let the life be temperate in every respect, and with a strong will the victory can be won." Alice B. Stockham, *Tokology: A Book for Every Woman* (New York: R. F. Fenno & Company, 1893), 160 (emphasis original).

[41] Rosalyn M. Meadow and Lillie Weiss, *Women's Conflicts about Eating and Sexuality: The Relationship Between Food and Sex* (New York and London: Routledge, 2012), 113.

[42] Trall, *Sexual Physiology and Hygiene*, 266.

Guidelines and rules on the diet were framed as ways to exercise self-restraint with food and drink, and as methods for the proper care of the self. Products for sexual inactivity or overactivity were not only advertised in publications; they were ***incorporated*** into marital sex advice literature. Product placement within advice literature connected guidelines on the revitalisation of a depleted sexual vigour with the consumption of nourishing remedies that could be obtained for a price. Readers were encouraged to act, first and foremost, as consumers in the pursuit of self-governance and self-improvement. A prominent example of this is Dr Frederick Hollick's *The Marriage Guide* (1860). In the subtitle, the work specifies that it is "a private instructor for married persons and those about to marry both male and female." The products targeting various ailments, including low sexual appetite, were firmly geared towards individuals who were or would be married. Hollick mentioned several treatments for men who have lost their "manly vigor": the use of a "hot stimulating lotion"[43] rubbed briskly on the penis to stimulate blood flow; and the use of a device called a "Congester," described as a pump into which the penis is inserted. While advising that many treatments can be used and consumed at home, Hollick reaffirmed the importance of medical expertise and advises that the Congester "is not an instrument adapted for self-treatment."[44] The patient-consumer was encouraged to govern their sexual appetites by availing themselves of possible curative devices, but they also needed to be monitored by the physician. Hollick even counselled flagellation and "firing"—which involved pressing a boiling-hot smooth iron button rapidly along the length of the penis.

Hollick was not completely clear in *The Marriage Guide* as to whether all the apparatuses were available for purchase from his practice, but he advertised an "aphrodisiac remedy" that could be obtained from him for $5. In an early example of direct-to-consumer advertising, the treatise featured the following announcement:

> it is not beyond all doubt the most generally and thoroughly efficacious remedy for impotence, sterility, loss of feeling, and natural torpidity, ever

[43] Frederick Hollick, *The Marriage Guide, or Natural History of Generation; A Private Instructor for Married Persons and Those about to Marry Both Male and Female; in Every Thing Concerning the Physiology and Relations of the Sexual System and the Production or Prevention of Offspring—Including All the New Discoveries, Never Before Given in the English Language* (New York: T W Strong, 1860), 149.

[44] Ibid., 150–157.

invented ... The medicine being in a dry form, there is no bottle needed, and therefore can be no risk of loss or breakage ... Remember, it can be obtained of no one else![45]

Low sexual appetite, in Hollick's configuration, was connected to energy more broadly. Men were suffering from a generalised sense of fatigue, which affected their sexual vigour and the health of the nation. Promising discretion and effectiveness, the marketing formula deployed by Hollick and other physicians combined direct-to-consumer advertising at the beginning of a manual that emphasised the importance of expert knowledge and customised medical advice for assuming individual responsibility for balancing one's appetite. Hollick's work demonstrates a notable shift from the publishing of advice on proper nutrition to the advertisements of chemical products in the forms of tablets. In a similar manner, but this time in the management of "nocturnal emissions," *The Philosophy of Marriage* (1862), by the doctors Jordan and Beck, provided the formula for "a very excellent diet drink" that consumers can prepare themselves.[46] In addition, they claimed that in some cases of impotence, "we have administered, with great effect, a lozenge (Morsuli)," for which they also published the formula.[47] The formulae can be obtained upon the doctors' receiving letters detailing the patient's condition and a "remittance of Five Dollars as consultation fee."[48] In the latter half of the nineteenth century, the purchasing power of the public gave way to a variety of possible products ranging from the mechanical and therapeutic to the chemical—in the shape of tablets and elixirs.

Patent Medicine and Reinvigorating the Appetites

Amidst advertising selling insurance, market produce and real estate in *The Omaha Daily Bee* in 1900, a doctor, F. G. Sanden from Chicago promised to cure men "without drugs" and offered instead a book and a "wonderful

[45] Hollick, *The Marriage Guide*, 30.

[46] Henry J. Jordan and Samuel Beck, *The Philosophy of Marriage Being Four Important Lectures, on the Function and Disorders of the Nervous System, and Reproductive Organs, Illustrated with Cases* (New York: Bloom & Smith, 1862), 113.

[47] Ibid., 115.

[48] Ibid., 173–174.

electric belt and suspensory for weak men."[49] In her history of the electric belt, Carolyn Thomas de la Peña argues that the mechanical device, which was claimed to improve sexual performance by "infusing the genitals with electric power," reflects a concern with manhood and the place of men in the modern world.[50] Electrotherapy, she writes, was a method to overcome the weaknesses of the body.[51] The electric belt, and the use of electricity more broadly, again reflected an enduring concern with the management of the body in late-nineteenth and early-twentieth centuries. While Thomas de la Peña argues that the use of electrotherapy was a method to master the body, she does not address how such discourses of mastery and equipment contributed to understandings of the management of sexual appetite.

Leafing through newspapers at the turn of the nineteenth century in North America, it is difficult to distinguish which advertisements were published by qualified physicians and which were so-called patent medicines. The nostrums were successfully marketed in the United States for centuries before the introduction of the Pure Food and Drug Act of 1906, which aimed to control the content and labelling of foods and drugs. The precursor to this Act was an investigative report by the journalist and "muckraker" Samuel Hopkins Adams titled *The Great American Fraud*, in which he exposed the plethora of false claims made by suppliers as well as the ingredients of some products that either did not cure patients or worsened their health.[52] At the turn of the century, Jane Marcellus remarks, the

[49] The Omaha Daily Bee, "30 Days' Trial: Dr. Sanden's Electric Belt," *The Omaha Daily Bee*, January 13, 1900, 6.

[50] Carolyn Thomas de la Peña, "Designing the Electric Body: Sexuality, Masculinity and the Electric Belt in America, 1880–1920," *Journal of Design History* 14, no. 4 (2001): 279. On harnessing electricity as a therapeutic tool, see Iwan Rhys Morus, "The Measure of Man: Technologizing the Victorian Body," *History of Science* 37, no. 3 (1999): 249–282.

[51] Thomas de la Peña, "Designing the Electric Body," 279. It is also worth noting that an "electric corset" emerged around the same time. The device claimed to treat issues such as women's weak nerves and hysteria. See Valerie Steele, *The Corset: A Cultural History* (New Haven: Yale University Press, 2001), 80–83.

[52] Samuel Hopkins Adams, *The Great American Fraud: Articles on the Nostrum Evil and Quacks* (P. F. Collier & Sons, 1905). For a discussion on the components of patent medicines, see J. Worth Estes, "The Pharmacology of Nineteenth-Century Patent Medicines," *Pharmacy in History* 30, no. 1 (1988): 3–18. "Muckraker" is a term coined by President Theodore Roosevelt to refer to writers who exposed the corruption of businesses or government to the public in the early-twentieth century. See Elizabeth Fee, "Samuel Hopkins Adams (1871–1958): Journalist and Muckraker," *American Journal of Public Health* 100, no. 8 (2010): 1390–1391.

advertising of patent medicine for the cure of lost manhood constituted a thriving industry, and alongside increased advertising, "growing distrust of orthodox medicine helped patent medicine use become even more widespread."[53] Demonstrating the extent of the lost manhood market, newspapers often featured ads from different suppliers on almost every page. On May 31, 1903, for example, *The Pittsburgh Press* published adverts for Dr Mackenzie, Dr McLaughlin, Dr Williams, F. G. Leslie M.D., Dr Morrell, Dr Koler, Dr Richardson, Dr Geo. A. Knox and Dr Ferris. The large number of advertisements demonstrates that low sexual appetite in men and associated ailments were a key source of revenue for the popular press.

Marketing to subjects involved appealing to their societal responsibilities as (re)productive citizens, and a balanced sexual appetite was inherent in this conceptualisation. Individuals were embroiled in commercial transactions and discourses of self-improvement. The variety of sources of information and products suggests that individuals were also encouraged to gather advice from a range of sources. The easy access to multiple doctors advertising their products and advice literature counselling on proper nourishment meant that subjects could assemble a range of commodities, whether written or ingested. In the following pages, I examine the invention, marketing and consumption of the aphrodisiac in more depth. I argue that techniques for the management of sexual appetites, combined with the marketing of devices and products, cultivated a subjectivity where the consumption of items was inextricable from the production of knowledge itself. In other words, the marketing and consumption of products encouraged individuals to consume and ingest knowledge, making such practices central to the formation of balanced sexual subjectivity.

Advertising in newspapers, marital sex advice literature and health treatises developed techniques of knowledge for the management of sexual appetite. These techniques were connected to consumer choices. Advice to individuals was not solely focused on how and what to ingest. It was also a commercial transaction that involved the advertisement of products for exercising mastery over sexual appetite. As impotence shifted to a physical problem or debility, an array of products in the form of elixirs and mechanical devices were developed and advertised. Doctors and quacks were interested in generating a profit from the widely circulated anxieties

[53] Marcellus, "Nervous Women and Noble Savages," 787.

over low sexual appetite and commercial products of varying content were offered as potential solutions.[54] As Kathleen L. Endres observes,

> By defining "Lost Manhood" as a physical condition that might be cured by pills, potions, lotions, pumps, straps, and firings, physicians, quacks, chemists, and the patent medicine industry were free to cash in, offering products that they claimed could cure the sexual dysfunction. In the process, the commercialization of impotence had begun in earnest.[55]

The kind of sexuality marketed to the public not only hinged on fears of lost manhood and perpetual fatigue, it also provided the promise of a controlled and youthful appetite, and a restoration of one's social position. Advertisements deployed a metaphor of usefulness to the nation. In other words, men were urged to master their bodily and nervous weaknesses to participate in the "pleasures and duties of life."[56] The reinvigoration of sexual appetite thus represented larger social issues. In the United States, the threat of a weakened nation and national identity was exemplified by the emergence of nervous diseases such as neurasthenia. The strengthening of sexual vigour would lead to a nation of "real" men, able to provide for their family, produce heirs and contribute to a thriving nation.[57]

Managing sexual appetite signified the cultivation of social stature and rank in society. The rejuvenation of appetites, as well as masculinity, status and hierarchy, were overarching themes in the marketing strategy. Advertisements found in newspapers at the time were particularly explicit about this risk. For instance, commercials addressed to "Weak Men in the Country" by the Wisconsin Medical Institute Physicians warned men to "master this weakness or lose your manhood." The Institute also attributed

[54] For a discussion on competition between quacks and physicians, and the deployment of the medical model of low sexual appetite, see chapter six in McLaren, *Impotence*, and James Harvey Young, "Patent Medicines: An Early Example of Competitive Marketing," *The Journal of Economic History* 20, no. 4 (1960): 648–656.

[55] Endres, "From 'Lost Manhood' to 'Erectile Dysfunction,'" 87.

[56] The Milwaukee Journal, "Weak Men in the Country," *The Milwaukee Journal*, December 10, 1904, 9.

[57] See chapter three in Michael S. Kimmel, *History of Men: Essays on the History of American and British Masculinities* (Ithaca, NY: State University of New York Press, 2005). Brett A. Berliner makes a similar point about France in "Mephistopheles and Monkeys: Rejuvenation, Race, and Sexuality in Popular Culture in Interwar France," *Journal of the History of Sexuality* 13, no. 3 (2004): 317.

the cause of such manly weaknesses to "early abuse or later excesses."[58] It offered readers a free 200-page book with "engravings and illustrations [advising] about the diseases of man."[59] Here, it is unclear whether the proposed product would be ingested in the form of an elixir or a tablet, or whether they were proposing the use of a mechanical device. However, manhood in this configuration entails sexual vigour and social functionality as well as the garnering of knowledge—consumerism meant access to knowledge on sexual appetite.

Marketing material promised a youthful, improved life to potential patients and consumers. The ads for mechanical and chemical products were not targeting the aged in particular; rather they generally traded in discourses of fear, fatigue and depletion of sexual vigour in men. Issues such as generalised debility and feeble nerves were recurrent in the selling of patent medicine.[60] Commercial notices harnessed principles of homeostasis, as the dominant ideas on health during the era relied on balance between different parts/organs of the body. Patent medicine drew on the idea that weakness in one part of the body could have detrimental, totalising effects for the self. As Charles Rosenberg notes, the body needed to maintain "its health-defining equilibrium."[61] Sexual appetite was included in this system: too much would draw energy from other parts of the body to the organ hence "leaving the depleted portions susceptible to disease."[62] Experiments and surgeries on endocrine glands would become especially prominent in Europe in the early-twentieth century.[63] In tandem with experiments in testicular transplantation in the early-twentieth century, the concern would shift to preserving (and replacing) the male hormone,

[58] The Milwaukee Journal, "Weak Men in the Country," 9.

[59] Ibid.

[60] Worth Estes, "The Pharmacology of Nineteenth-Century Patent Medicines," 4.

[61] Charles E. Rosenberg, "The Therapeutic Revolution: Medicine, Meaning, and Social Change in Nineteenth-Century America," *Perspectives in Biology and Medicine* 20, no. 4 (1977): 495.

[62] Gail Pat Parsons, "Equal Treatment for All: American Medical Remedies for Male Sexual Problems: 1850–1900," *Journal of the History of Medicine and Allied Sciences* 32, no. 1 (1977): 59.

[63] See Chandak Sengoopta, "'Dr Steinach coming to make old young!': Sex Glands, Vasectomy and the Quest for Rejuvenation in the Roaring Twenties," *Endeavour* 27, no. 3 (2003): 122–126, also by Sengoopta, "Glandular Politics: Experimental Biology, Clinical Medicine, and Homosexual Emancipation in Fin-de-Siècle Central Europe," *Isis* 89, no. 3 (1998): 445–473, and Brett A. Berliner, "Mephistopheles and Monkeys." See also chapter five in Nikolai Krementsov, *Revolutionary Experiments: The Quest for Immortality in Bolshevik Science and Fiction* (New York: Oxford University Press, 2014).

which one could not control. Such discourses formed the historical ground for the development of oestrogen replacement therapy (or hormone replacement therapy). As Elizabeth Siegel Watkins writes, "[A] consequence of this new reasoning meant that men did not have to conserve their semen by avoiding ejaculation."[64] In the late-nineteenth century, the concern with the restoration of vitality, associated with sexual appetite, was prevalent in the popular and medical press.

A rejuvenated, balanced sexual appetite signified comprehensive bodily equilibrium. The body was approached in the American popular press as a total entity where the balance between the parts signified the health of the whole. Advertising of products for lost manhood reflected this idea by openly promising panaceas that would cure all ailments. The Hallock Medical Institute in Boston, for example, promised to cure "all Diseases and Weaknesses of Man, from whatever cause permanently and privately ... by the use of The Old Dr. Hallock Electric Pills." The Institute boasted that pills, at $1 a box and sent by mail, have been used "since 1848, and with universal success."[65] The selling of universal remedies promised to fight any weaknesses and return "feeling" to individuals. The advertising regularly drew on the language of "sensation"—individuals would be able to gain strength and affective qualities and function in society confidently. The senses were harnessed in *fin-de-siècle* society as relational, bodily, and marketable. We see a confluence of motifs in the selling of products for sexual appetite: equilibrium between body parts, rejuvenation of vigour (itself irreducible to seminal fluid) and renewal of affective sensations. These tropes buttressed the development of a consumerist subjectivity. As Michael S. Kimmel reminds us, historians repeatedly note that "the turn of the century [was] an era of transition from a 'culture of production' to a 'culture of consumption'."[66]

[64] Elizabeth Siegel Watkins, *The Estrogen Elixir: A History of Hormone Replacement Therapy in America* (Baltimore: The Johns Hopkins University Press, 2007).

[65] Lewiston Evening Journal, "Manly Vigor," *Lewiston Evening Journal*, October 12, 1897, 6.

[66] Kimmel, *History of Men*, 43.

ENGAGING THE "MANLY VIGOUR" OF THE CONSUMER

In his study of the "therapeutic roots" of consumer culture from 1880 to 1930, the historian T.J. Jackson Lears observes that during the late-nineteenth and early-twentieth centuries, "the advertisers' audience was neither as passive nor as gullible as critics sometimes assumed."[67] Rather the advertising of a range of products, including those for the treatment of "manly vigor," formed the convergence of discourses of disease, therapeutic goals and consumerism. Consumers of products for the governance of sexual appetite emerge as participative and complicit in the products they utilise and in the health of the nation as a whole, which was intertwined with the appeal to consumerism: "Promising wholeness or rejuvenation, advertisers addressed those immersed in routine work *or* domestic drudgery; they held out the hope that life could be perpetually fulfilling; and they implied that one ought to strive for that fulfillment through consumption."[68]

This approach to advertising allows for an examination of processes whereby medical expertise on the fluctuations of sexual appetite, popular patent medicine and widespread accessibility of materials—to be read and to be ingested—coalesce to abet and buttress consumerist desires as well as individuals' social circumstances, and participate in the formation of sexual subjectivities. The discourse of the popular press resonates with the people, and it also constitutes a site of contestation where meaning is unstable and renegotiated. Individuals, then, are always embroiled in cultivating oscillating meanings. The distribution of images, signs and discourses are always characterised by an ambivalent consumption, and audiences derive and reproduce diverse layers of meanings from such consumptive practices.

The adverts that have been discussed in this chapter deployed terms that would resonate with future consumers, made use of (real or fictional) case studies, cited statistics of successful treatment and referred to consumer reviews, where they were often portrayed as grateful and humbled by the treatment.[69] Yet individuals in the late-nineteenth and early-twentieth

[67] T.J. Jackson Lears, "From Salvation to Self-Realization: Advertising and the Therapeutic Roots of the Consumer Culture, 1880–1930," in *The Culture of Consumption: Critical Essays in American History, 1880–1980*, eds. Richard Wightman Fox and T.J. Jackson Lears (New York: Pantheon Books, 1983), 28.

[68] Ibid., 27 (emphasis original).

[69] Young, "Patent Medicines," 654.

centuries were not blindly seduced by the marketing of products for "lost manhood," particularly given that they were presented with a wide variety of choices from a large number of medical or non-medical sources. Certainly, advertising harnessed socio-sexual concerns, but individuals were not passive to ideological messages. Presented with a range of messages from doctors and quacks, patient-consumers actively participated in a process of decoding messages, negotiating knowledge and creating meaning. In Canada, for example, "The Dr. Williams' Medicine Company" from Ontario marketed "Dr. Williams' Pink Pills." The marketing of those pills addressed a "universal" experience of masculinity: "Has it ever occurred to you that you need a medicine as men—not as old men or young men, but as men? Are you never conscious that the special wear and tear of life which men sustain need repair?"[70] Dr Williams' Pink Pills "restore manly vigor and energy" and towards the end of the advert there is a quick mention of "women, too."[71] The appeal to a common masculinity in need of healing alongside the active nature of consumers transcended geographical boundaries. These examples highlight how the turn of the century marked a transformation in consumer culture and in marketing techniques.

The marketing of elixirs for managing manly vigour was adapted to its audience in the nineteenth and twentieth centuries and was mobilised to produce normative frameworks of manhood, usefulness and social reputation. And central to the consumption of such advertisements was the body as a vehicle for balanced sexual appetite. For a cultural product to be successful, it must understand its market and underscore a social need or concern. The rapid rise of the mass media and networks of communication threatened to overwhelm consumers with too many messages and flood the market with uncertified or untested inventions. This resulted in parts of the (pseudo-)medical profession to simplify their offerings, but also educate consumers about their consumptive capacities and abilities.

For example, the so-called Union Physicians from Pittsburgh, Pennsylvania, offered a "Pelvic method" to men suffering from lost manhood. They were also cautious to reassure prospective consumers that they would not receive junk mail: "[everyone] who writes to us may feel assured that they will receive no mail from us except in answer to theirs."

[70] The Montreal Gazette, "Dr. Williams' Pink Pills," *The Montreal Gazette*, March 16, 1904, 2.

[71] Ibid.

Unsolicited books and questionnaires were a characteristic of "unreliable doctors and medical institutions."[72] In contrast to Knelman, who argues that understandings of disease were "absorbed by anyone who read the papers, which, because they were expensive, were perused thoroughly,"[73] I argue that far from passive pawns in a competitive market, consumers were urged to inform themselves and to learn how to discriminate between genuine qualified physicians and quacks. This distinction was often unclear. However, they overwhelmingly tended to communicate a need for consumers to inform themselves both on sexual appetite and on lurking diseases. The printed material demonstrates an attention to education and information. It does more than position "male sexuality as a matter of will."[74] It also participates in the formation of a balanced sexual subjectivity. In addition, while expertise was certainly a marketing technique, the management of sexual appetite (and its commerce) relied on an informed audience. A conspicuous ad in the Pittsburgh Press by Dr Kane's New York Medical Specialist Company asked men: "What is Blood Poison? Read!" "What is Lost Manhood? Read!" "What is Varicocele? Read!" and "What is Stricture? Read!" Dr Kane's company claimed "electricity and absorption" as the "only sure method." Dr Kane's ad appealed to all men offering to "do more for you for $5.00 than any other doctor will for $25.00 and providing care for the poor as well as the rich [...] My motto is 'Live and Let Live.'"[75]

What connects the transformation from a "culture of production" to a "culture of consumption" at the turn of the century is a preoccupation with the body and balance in the domain of sexuality. As Kimmel observes, by the 1830s, a "marketplace [for] manhood" emerged in North America. Masculine identity was derived from participation in capitalist systems of consumption and ownership. However, a prerequisite for success in the marketplace was also bodily control. Bodily (and sexual) balance thus was a major feature in tracts and advertising. Gaining mastery over sexual appetite meant wealth and social success. By the end of the nineteenth century, "industrial capitalism [would require] adventurous consumers."[76]

[72] The Pittsburgh Press, "Reliable Cures by True Specialists," *The Pittsburgh Press*, October 17, 1903, 4.

[73] Judith Knelman, "Nervous Debility: A Disorder Made to Order," *Victorian Review* 22, no. 1 (1996): 35.

[74] Ibid., 39.

[75] The Pittsburgh Press, "Weak Diseased Men," *The Pittsburgh Press*, May 31, 1903, 18.

[76] Kimmel, *History of Men*, 49.

Suppliers also demonstrated extensive marketing savvy, with one ad offering to refund the train journey of clients: "Every train brings some man from a distance to be cured. Railroad fare deducted for out-of-town patients coming for an examination."[77] Customers were encouraged to travel far and wide to access tools to govern their sexual appetites.

The advertising of products for the management of sexual appetite reveal not only an enduring concern about male sexual appetite and performance, it also sheds light on the capitalist competition that occurred. Consumption and the recovery of manhood were amalgamated at the turn of the century as quacks and physicians competed for a share of the market of socio-sexual "problems" by introducing an array of elixirs and mechanical devices to the marketplace. Harnessing the vocabulary of rejuvenation and self-improvement, advertising encouraged individuals to invest time and energy in their bodies. The large number of products and sources that could be used by consumers reflects the perpetuation of diseases of manly vigour.

From nourishment to herbal remedies to mechanical devices, the management of sexual appetite was a recurrent concern in late-nineteenth and early-twentieth century North America. The concern with the potency of foods and drinks emerged in medical and non-specialist writings and was embedded within an overall consideration of familial and conjugal relations. Sexual appetite was problematised as an indicator of morality and stability both for the family unit and for the nation more broadly. Individuals were urged to develop skills to triumph over their baser instincts, which were encapsulated in the recurrent problematisation of "how much sex" and "how frequent" sexual relations should be. Balance and order were necessary features of civilised societies, and the threat of sexual decadence was ever-present. Inhibition signified control rather than complete sublimation, given that an amount of sexual appetite was necessary for generation. Drawing on the language of regulation and homeostasis, individuals were advised to monitor their nutrition because all organs in the body needed a state of balance for health. While the publication of *Onania* by Samuel-Auguste Tissot in 1760 had certainly popularised the idea that the intensity of sexual appetite is influenced by diet,[78] the analysis of plants, roots and so on for their effects on sexual appetite can be traced to the Ancient Greek physician and botanist Pedanius Dioscorides

[77] The St Paul Globe, "Lost Manhood," *The St Paul Globe*, August 18, 1903, 6.
[78] Nissenbaum, *Sex, Diet, and Debility in Jacksonian America*, 33–34.

(c. 40–90 CE) who published his findings in *De Materia Medica* (originally written c. 50–70 CE).[79] Although the concern with dietary regimes and sexual appetite does not originate in nineteenth-century North America, what this chapter has shown is how that era witnessed an expansion of techniques for managing sexual appetite with dietary regimes and chemical and mechanical tools. At the same time, the flourishing of newspaper and advertising industries, which in turn conditioned the emergence of an active consumer, accompanied the production of discourses on appropriate regimen.

While nutrition continued to preoccupy writers, in the nineteenth century, the booming newspaper business gave rise to the publication of all manners of elixirs, powders and mechanical devices for the treatment of "lost manhood." Impotence continued to figure in those ads as a moral condition resulting from excessive sexual appetite, masturbation and spermatorrhoea. However, the treatment was a mixture of ingestible products and elixirs, devices such as pumps or even actions on the penis such as flagellation. The patent medicine industry flourished at this time by drawing on discourses of hope, renewal and rejuvenation. They promised both youthful potency and control over sexual appetite to consumers. Ideas on masculinity were intimately linked to commercial success to "escape the civilizing constraints of domestic life represented by the Victorian woman."[80] As Kimmel notes, "the self-control required of marketplace success required the sexual control of a disciplined body, a body controlled by the will."[81] The widespread adverts promised control over the problem of lost manhood. The ads developed a narrative that hinged on ideas of performance, production and self-improvement. Indeed, consumers were encouraged to read and inform themselves on different diseases, not solely connected to sexual appetite. What we witness, then, is that the medicalisation of sexual appetite took place during this era within a system of consumerism and the circulation of commodities.

This chapter has examined the commodification of sexual appetite through the advertisements of elixirs and mechanical devices in manuals, pamphlets, tracts and newspapers in the late-nineteenth and early-twentieth

[79] See Pedanius Dioscorides, *De Materia Medica: Being an Herbal with Many Other Medicinal Materials Written in Greek in the First Century of the Common Era*, trans. Tess Anne Osbaldeston (Johannesburg: Ibidis, 2000).

[80] Kimmel, *The History of Men*, 39.

[81] Ibid., 40.

centuries in the United States. Following the end of the Second World War and the movement of sexologists and psychiatrists from Europe, the United States witnessed a transformation from an informal industry of "quacks" marketing aphrodisiacs in the pursuit of reviving manly vigour to the emergence of the modern psychiatric institution. In the next chapter, the development of psychiatric knowledge on sexual appetite in the twentieth century is examined through two routes. First, the chapter examines the work of the team led by Alfred Charles Kinsey in the canvassing of interviews to produce statistics and averages on human sexuality. In these representations, sexual appetite was to be counted and mapped. Chapter 4 then analyses the research of William H. Masters and Virginia E. Johnson, in particular their use of tools for studying human sexual response. Their work cemented norms of sexual appetite, presenting both the necessity of perfecting techniques to achieve pleasure and the norm to which individuals should aspire. This chapter contends that the works of Kinsey, and Masters and Johnson were important for reifying concepts of averages and norms and for developing techniques for the measurement of sexual appetite.

Bibliography

Bayuk Rosenman, Ellen. 2003. Body Doubles: The Spermatorrhea Panic. *Journal of the History of Sexuality* 12 (3): 365–399.

Beard, George Miller. 1881. *American Nervousness: Its Causes and Consequences, a Supplement to Nervous Exhaustion (Neurasthenia)*. New York: G. P. Putnam's Sons.

———. 1884. *Sexual Neurasthenia: Its Hygiene, Causes, Symptoms, and Treatment with a Chapter on Diet for the Nervous*. New York: E. B. Treat.

Berliner, Brett A. 2004. Mephistopheles and Monkeys: Rejuvenation, Race, and Sexuality in Popular Culture in Interwar France. *Journal of the History of Sexuality* 13 (3): 306–325.

Buringh, Eltjo, and Jan Luiten van Zanden. 2009. Charting the 'Rise of the West': Manuscript and Printed Books in Europe, a Long-Term Perspective from the Sixth Through Eighteenth Centuries. *The Journal of Economic History* 69 (2): 409–445.

Cott, Nancy F. 1978. Passionlessness: An Interpretation of Victorian Sexual Ideology. *Signs: Journal of Women in Culture and Society* 4 (2): 219–236.

Cryle, Peter, and Alison Moore. 2011. *Frigidity: An Intellectual History*. London: Palgrave Macmillan.

D'Emilio, John, and Estelle B. Freedman. 2012 [1988]. *Intimate Matters: A History of Sexuality in America*. 3rd ed. New York: Harper & Row.

Dean, Carolyn J. 1996. *Sexuality and Modern Western Culture*. New York: Twayne Publishers.

Dioscorides, Pedanius. 2000. *De Materia Medica: Being an Herbal with Many Other Medicinal Materials Written in Greek in the First Century of the Common Era*. Trans. Tess Anne Osbaldeston. Johannesburg: Ibidis.

Eisenstein, Elizabeth. 1979. *The Printing Press as an Agent of Change*. Cambridge: Cambridge University Press.

Endres, Kathleen L. 2014. From 'Lost Manhood' to 'Erectile Dysfunction': The Commercialization of Impotence. In *We Are What We Sell: How Advertising Shapes American Life…and Always Has, Volume 2: Advertising at the Center of Popular Culture: 1930s–1975*, eds. Danielle Sarver Coombs and Bob Batchelor, 84–100. Santa Barbara, CA: Praeger.

Evans, Jennifer. 2014. *Aphrodisiacs, Fertility and Medicine in Early Modern England*. Suffolk: The Boydell Press.

Fee, Elizabeth. 2010. Samuel Hopkins Adams (1871–1958): Journalist and Muckraker. *American Journal of Public Health* 100 (8): 1390–1391.

Fowler, L.N. 1848. *Marriage: Its History and Ceremonies; with a Phrenological and Physiological Exposition of the Functions and Qualifications for Happy Marriages*. New York: Fowler & Wells.

Gabler, William G. 1978. The Evolution of American Advertising in the Nineteenth Century. *The Journal of Popular Culture* XI (4): 763–771.

Gordon, Michael. 1969. The Ideal Husband as Depicted in the Nineteenth Century Marriage Manual. *The Family Coordinator* 18 (3): 226–231.

———. 1971. From an Unfortunate Necessity to a Cult of Mutual Orgasm: Sex in American Marital Education Literature 1830–1940. In *Studies in the Sociology of Sex*, ed. James M. Henslin, 53–77. New York: Meredith Corporation.

Gordon, Michael, and M. Charles Bernstein. 1970. Mate Choice and Domestic Life in the Nineteenth-Century Marriage Manual. *Journal of Marriage and Family* 32 (4): 665–674.

Hollick, Frederick. 1860. *The Marriage Guide, or Natural History of Generation; A Private Instructor for Married Persons and Those About to Marry Both Male and Female; in Every Thing Concerning the Physiology and Relations of the Sexual System and the Production or Prevention of Offspring—Including All the New Discoveries, Never Before Given in the English Language*. New York: T W Strong.

Hopkins Adams, Samuel. 1905. *The Great American Fraud: Articles on the Nostrum Evil and Quacks*. P. F. Collier & Sons.

Jefferis, B.G., and J.L. Nichols. 1894. *Search Lights on Health: Light on Dark Corners, A Guide to Purity and Physical Manhood. Advice to Maiden, Wife and Mother. Love, Courtship and Marriage*. Canada: The J L Nichols Company.

Jordan, Henry J., and Samuel Beck. 1862. *The Philosophy of Marriage Being Four Important Lectures, on the Function and Disorders of the Nervous System, and Reproductive Organs, Illustrated with Cases.* New York: Bloom & Smith.

Kellogg, John Harvey. 1877. *Plain Facts About Sexual Life.* Battle Creek, MI: Office of the Health Reformer.

Kimmel, Michael S. 2005. *History of Men: Essays on the History of American and British Masculinities.* Ithaca, NY: State University of New York Press.

Knelman, Judith. 1996. Nervous Debility: A Disorder Made to Order. *Victorian Review* 22 (1): 32–41.

Krementsov, Nikolai. 2014. *Revolutionary Experiments: The Quest for Immortality in Bolshevik Science and Fiction.* New York: Oxford University Press.

Lallemand, Claude-François. 1861 [1840]. *A Practical Treatise on the Causes, Symptoms and Treatment of Spermatorrhoea.* Trans. Henry J. McDougall. Philadelphia: Blanchard and Lea.

Lears, T.J. Jackson. 1983. From Salvation to Self-Realisation: Advertising and the Therapeutic Roots of the Consumer Culture, 1880–1930. In *The Culture of Consumption: Critical Essays in American History, 1880–1980*, eds. Richard Wightman Fox and T.J. Jackson Lears, 1–38. New York: Pantheon Books.

Lewiston Evening Journal. 1897. Manly Vigor. *Lewiston Evening Journal*, October 12, p. 6.

Marcellus, Jane. 2008. Nervous Women and Noble Savages: The Romanticized 'Other' in Nineteenth Century US Patent Medicine Advertising. *The Journal of Popular Culture* 41 (5): 784–808.

McLaren, Angus. 2007. *Impotence: A Cultural History.* Chicago and London: University of Chicago Press.

Meadow, Rosalyn M., and Lillie Weiss. 2012. *Women's Conflicts About Eating and Sexuality: The Relationship Between Food and Sex.* New York and London: Routledge.

Melody, M.E., and Linda M. Peterson. 1999. *Teaching America About Sex: Marriage Guides and Sex Manuals from the Late Victorians to Dr. Ruth.* New York: New York University Press.

Morus, Iwan Rhys. 1999. The Measure of Man: Technologizing the Victorian Body. *History of Science* 37 (3): 249–282.

Mumford, Kevin J. 1992. 'Lost Manhood' Found: Male Sexual Impotence and Victorian Culture in the United States. *Journal of the History of Sexuality* 3 (1): 33–57.

Neuhaus, Jessamyn. 2000. The Importance of Being Orgasmic: Sexuality, Gender and Marital Sex Manuals in the United States, 1920–1963. *Journal of the History of Sexuality* 9 (4): 447–473.

Nissenbaum, Stephen. 1980. *Sex, Diet, and Debility in Jacksonian America: Sylvester Graham and Health Reform.* Westport, CT: Greenwood Press.

Parsons, Gail Pat. 1977. Equal Treatment for All: American Medical Remedies for Male Sexual Problems: 1850–1900. *Journal of the History of Medicine and Allied Sciences* 32 (1): 55–71.
Rosenberg, Charles E. 1977. The Therapeutic Revolution: Medicine, Meaning, and Social Change in Nineteenth-Century America. *Perspectives in Biology and Medicine* 20 (4): 485–506.
Ryan, Michael. 1837. *The Philosophy of Marriage, in Its Social, Moral, and Physical Relations*. London: John Churchill.
Sengoopta, Chandak. 1998. Glandular Politics: Experimental Biology, Clinical Medicine, and Homosexual Emancipation in Fin-de-Siècle Central Europe. *Isis* 89 (3): 445–473.
———. 2003. 'Dr Steinach coming to make old young!': Sex Glands, Vasectomy and the Quest for Rejuvenation in the Roaring Twenties. *Endeavour* 27 (3): 122–126.
Smith-Rosenberg, Carroll. 1985. *Disorderly Conduct: Visions of Gender in Victorian America*. New York: Oxford University Press.
Steele, Valerie. 2001. *The Corset: A Cultural History*. New Haven: Yale University Press.
Stephens, Elizabeth. 2008. Pathologizing Leaky Male Bodies: Spermatorrhea in Nineteenth-Century British Medicine and Popular Anatomical Museums. *Journal of the History of Sexuality* 17 (3): 421–438.
Stockham, Alice B. 1893. *Tokology: A Book for Every Woman*. New York: R F Fenno & Company.
The Milwaukee Journal. 1904. Weak Men in the Country. *The Milwaukee Journal*, December 10, p. 9.
The Montreal Gazette. 1904. Dr. Williams' Pink Pill. *The Montreal Gazette*, March 16, p. 2.
The Omaha Daily Bee. 1900. 30 Days' Trial: Dr. Sanden's Electric Belt. *The Omaha Daily Bee*, January 13, p. 6.
The Pittsburgh Press. 1903. Reliable Cures by True Specialists. *The Pittsburgh Press*, October 17, p. 4.
———. 1903. Weak Diseased Men. *The Pittsburgh Press*, May 31, p. 18.
The St Paul Globe. 1903. Lost Manhood. *The St Paul Globe*, August 18, p. 6.
Thomas de la Peña, Carolyn. 2001. Designing the Electric Body: Sexuality, Masculinity and the Electric Belt in America, 1880–1920. *Journal of Design History* 14 (4): 275–289.
Trall, R.T. 1891. *Sexual Physiology and Hygiene: An Exposition Practical, Scientific, Moral, and Popular, of Some of the Fundamental Problems in Sociology*. New York: M. L. Holbrook & Company.
Walters, Ronald G. 2000. *Primers for Prudery: Sexual Advice to Victorian America*. Baltimore, MD: Johns Hopkins University Press.

Watkins, Elizabeth Siegel. 2007. *The Estrogen Elixir: A History of Hormone Replacement Therapy in America*. Baltimore: The Johns Hopkins University Press.

Worth Estes, J. 1988. The Pharmacology of Nineteenth-Century Patent Medicines. *Pharmacy in History* 30 (1): 3–18.

Wrobel, Arthur, ed. 2015. *Pseudo-Science and Society in 19th-Century America*. Lexington: The University Press of Kentucky.

Young, James Harvey. 1960. Patent Medicines: An Early Example of Competitive Marketing. *Journal of Economic History* 20 (4): 648–656.

CHAPTER 4

Measuring Sex

On January 2, 1950, *The Eugene Register-Guard*, a daily newspaper based in Eugene, Oregon, published an article titled "Normal Sexual Behavior? No Such Thing Scientists." The article reports on scientists' responses to the publication of *Sexual Behavior in the Human Male* (1948) and notes their belief that practices found in the report should be treated as normal. In the article, the anthropologist Ralph Linton declares that there is no such thing as normal sexual behaviour "aside from the standards of normalcy set by the individual's society." Linton goes as far as asserting that the actions reported in fact "'fall far short of the actual range of sex behaviour', in humans, and 'by no means cover the possibilities.'"[1] This article is amongst the many published commentaries on the reception of *Sexual Behavior in the Human Male* in North America.

Research in sexuality in the nineteenth and early twentieth centuries predominantly relied on patients' narratives in the form of case histories. Alfred Charles Kinsey's studies however broke with this tradition by coding interviewee responses into statistical data. Such an approach had many detractors and Donna J. Drucker attributes the critiques of Kinsey's methodology as reflecting a concern with "the shift from case history models to statistical models."[2] By the 1960s, William H. Masters and Virginia

[1] The Eugene Register-Guard, "Normal Sexual Behavior? No Such Thing Scientists," *The Eugene Register-Guard*, January 2, 1950, 2A.

[2] Donna J. Drucker, "'A Most Interesting Chapter in the History of Science': Intellectual Reponses to Alfred Kinsey's *Sexual Behavior in the Human Male*," *History of the Human Sciences* 25, no. 1 (2012): 84.

J. Flore, *A Genealogy of Appetite in the Sexual Sciences*,
https://doi.org/10.1007/978-3-030-39423-3_4

E. Johnson would further expand the turn to tools of measurement in the study of human sexuality.

The measurement of normality, and its role in the conceptualisation of sexual appetite, is the central focus of this chapter, which is divided into two sections that respectively examine the works of Kinsey, and Masters and Johnson, in the development of sexual science in the twentieth century.[3] A genealogy of sexual appetite must attend to the techniques of calculation and observation that are central to the works of Kinsey, Masters and Johnson. The chapter thus analyses how Kinsey utilised statistics and the concept of averages in his research on human sexuality. The mapping of human sexuality through the compilation of data would, according to Kinsey and his team, offer an objective image of sexuality. In mid-twentieth-century North America, psychoanalysis still dominated knowledge of sexuality. Explanations for sexual shortcomings relied on neuroses, sublimation and repressed desires; few researchers interviewed patients to gather numerical patterns, constellations or paradigms. Kinsey, Janice Irvine writes, "felt that the success of his research depended on scientific rigor and … on his ability to convince the public of the stringency and objectivity of his approach."[4] While the methodology of the Kinsey studies was not completely unprecedented, the large number of research participants[5] alongside the wide availability of the results contributed to passionate responses to the Kinsey reports. This chapter argues that sexual appetite conditioned how statistical data was used in the Kinsey studies. The Kinsey team mobilised questions of "how much?" and "how often?" to produce graphs on which sexual appetite could be counted and mapped.

Turning to the work of Masters and Johnson, specifically their 1966 book *Human Sexual Response*, the chapter explores how the researchers further opened sexual activity to scientific investigation. Masters and Johnson designed the human sexual response cycle (HSRC), comprising excitement, desire, plateau and orgasm, which would go on to define future scientific studies of sexuality and, importantly, the "Psychosexual

[3] For a detailed historical study of the concept of normal, see Peter Cryle and Elizabeth Stephens, *Normality: A Critical Genealogy* (Chicago: University of Chicago Press, 2017).

[4] Janice Irvine, *Disorders of Desire: Sexuality and Gender in Modern American Sexology* (Philadelphia: Temple University Press, 2005), 22.

[5] Kinsey has been criticised for his lack of inclusion of racial and ethnic diversity in terms of both researchers (male, heterosexual and white Anglo-Saxon Protestants) and his research subjects. See Irvine, *Disorders of Desire*, 25.

Disorders" in the third edition of the *Diagnostic and Statistical Manual of Mental Disorders* (1980). Their work cemented norms of sexual appetite, presenting both the necessity of perfecting tools for achieving sexual pleasure and the norm to which individuals should aspire. This chapter thus contends that the works of Kinsey, Masters and Johnson were important for reifying concepts of averages and norms, but also for developing techniques for the observation of sexual appetite. What emerges from their research methodology is a calculable subject whose sexual appetites can be counted and mapped against an average, and whose physiology can be observed and compared against a norm.

Early-Twentieth-Century Sex Studies

Before opening Kinsey's reports and the works of Masters and Johnson, this chapter first explores how questions of amount featured in early-twentieth-century research on sex. The interwar years in North America were marked by significant social, cultural and political changes. The status of women, in particular, underwent important transformations. World War I bolstered the presence of women in the job market, women obtained the right to vote in 1920 and discussions on birth control widened significantly. As John D'Emilio and Estelle B. Freedman write, the United States, in the 1920s, was moving towards "an overlapping set of beliefs that detached sexual activity from the instrumental goal of procreation, affirmed heterosexual pleasure as a value in itself, defined sexual satisfaction as a critical component of personal happiness and successful marriage."[6] Renewed interest in sexuality was partly fuelled by the ideas of Sigmund Freud and Havelock Ellis, whose works were already influential in North America since the late nineteenth century.[7] In addition, sexual pleasure increasingly became understood as an important component of a fulfilling marriage, placing reproduction as a significant but somewhat secondary endeavour. Female sexual pleasure especially became synonymous with adventure and danger as revealed by growing discussions on contraception and abortion through the early mobilisation of the birth control movement.[8]

[6] John D'Emilio and Estelle B. Freedman, *Intimate Matters: A History of Sexuality in America*, 3rd ed (New York: Harper & Row, 2012 [1988]), 241.

[7] Ibid., 223–225.

[8] Ibid., 242–243.

From the 1920s to the 1950s, physicians expressed concern with the stability of the institution of marriage. Theodore Van de Velde's influential *Ideal Marriage: Its Physiology and Technique*, published in England in 1926, asserted the centrality of sexual pleasure to maintaining happiness in marriage. Van de Velde, a Dutch gynaecologist, explained that proper education of men and women on sexuality and pleasure was necessary to strengthen marriage. He argued that the cornerstones of the "temple of love and happiness in marriage" included "a vigorous and harmonious sex life."[9] Sex reformers and researchers continued to gather and publish data from studies on attitudes towards birth control, sex education and marriage during the first half of the century. By the 1950s, surveys and public opinion polls on diverse topics were becoming common in American society; hence, surveys on sexuality were not unheard of before Kinsey. One of the key precursors to Kinsey's research in the United States was Katharine Bement Davis' study, *Factors in the Sex Life of Twenty-Two Hundred Women* (1929). The topics covered by Davis were extensive and provided a framework for future surveys on sexuality. Her questionnaire delved into topics such as childhood, menstruation, adolescence and marriage, and paid attention to differences in education and attitudes towards sexuality amongst women. The study also asked questions about masturbation, contraception, frequency of sexual desire, and homosexuality.[10]

In *An American Obsession*, Jennifer Terry examines Davis' influence on scientific sex research almost exclusively in terms of what it revealed about homosexuality. As Terry observes, there were no "intimations of pathology, inversion, or constitutional difference" in Davis' study.[11] Thus, one of the most important conclusions of Davis' work, for Terry—especially in relation to its influence on subsequent sex research—was uncovering that "homoerotic and autoerotic experiences were common"[12] in the lives of women. However, what is also highly significant for the purposes of this chapter was how Davis devoted part of her book to the "periodicity of sex

[9] Theodore Hendrik Van de Velde, *Ideal Marriage: Its Physiology and Technique* (London: William Heinemann, 1940 [1926]), 2.

[10] Katharine Bement Davis, *Factors in the Sex Life of Twenty-Two Hundred Women* (New York: Harper & Brothers, 1929). See also chapter four in Jennifer Terry, *An American Obsession: Science, Medicine, and Homosexuality in Modern Society* (Chicago and London: Chicago University Press, 1999).

[11] Terry, *An American Obsession*, 131.

[12] Ibid., 134.

desire."[13] This attention to fluctuations in desire suggests an awareness of amount and frequency in sexuality. Indeed, women were asked about "sex feeling" and whether they can identify times, often related to menstruation, when they experience more or fewer difficulties. While this might point to the question of reproduction, it also reveals how the measurement of sexual appetite was important for sex research in the twentieth century.

The rise of women's movements, the subject of women's satisfaction, sexual or otherwise, and a general concern with the endurance of the institution of marriage were central to sex research throughout the first half of the twentieth century.[14] Gilbert V. Hamilton's *A Research in Marriage* (1929) examined the lives of two hundred married men and women. Hamilton, a psychoanalyst, asked questions about childhood, family life, relationships, opinions on sex and domestic life. Similar to Davis' study, Hamilton devised a range of questions on fluctuations in sexual appetite. Questions included: "What effect, if any, did pregnancy have on your sex desire? Did it increase or decrease it?", "Did any of these operations affect your sex desire?" and one measuring the "Relation of periodicity of sex desire to orgasm capacity."[15] Answers to questions were then added and percentages extracted. The technique of measurement, at this point, was less about letting patients freely speak about their sexual history, and more focused on guided question-and-answer sessions. The technique of calculation in Hamilton's work has been discussed in terms of homosexual object choice. Several works on the history of sexuality, while recognising the importance of Hamilton's contribution, do not examine his attention to sexual appetite or how he discussed sexual appetite in relation to homosexuality.[16] This is revealing of the constitutive relationship between sexual appetite and object choice in the genealogy of sexuality.

Another research project, sponsored by the Committee for the Study of Sex Variants (CSSV) in the mid-1930s, was an important precursor of the

[13] See chapters eight and nine in Davis, *Factors in the Sex Life*.

[14] Terry, *An American Obsession*, 129.

[15] Gilbert V. Hamilton, *A Research in Marriage* (New York: Lear, 1948 [1929]), 129, 137, 197.

[16] Examples include Erin G. Carlston, "'A Finer Differentiation': Female Homosexuality and the American Medical Community, 1926–1940," in *Science and Homosexualities*, ed. Vernon A. Rosario (New York: Routledge, 1997), 177–196, Lillian Faderman, *Odd Girls and Twilight Lovers: A History of Lesbian Life in Twentieth-Century America* (New York: Columbia University Press, 2012 [1991]), and Terry, *An American Obsession*.

Kinsey studies. Although the committee was dedicated to the study of homosexuality, their deployment of the term "sex variance" indicated a move "toward a more fluid paradigm that was based on statistical averages."[17] The significance of statistics and ideas of a continuum in sexuality were already circulating in research on homosexuality by the 1930s. The CSSV study combined nineteenth-century ideas that were well-known in psychiatry—for example, arrested psychosexual development, atavism and hereditariness—and asked participants a range of questions on their socio-economic background, occupation, and included photographic material of naked bodies as well as extensive physiological examinations. The researchers devoted a large part of their study on their participants' anatomy. This was not confined to the study of genitalia, as shoulders, abdomens, facial contours, hair distribution and voices were also examined for supposed signs of homosexuality. The study of homosexuality by the committee was guided by a will to strengthen heterosexuality and marriage, and prevent a "spread" of homosexuality in American society through enforcing "proper reproduction."[18] Although focused on object choice, much like Davis' and Hamilton's work, the *Sex Variants* study collected interviewees' assessments on the frequency of their sexual relations as well as "sex desire."[19] This example reveals the contingencies of sexual appetite and object choice, whereby frequency and desire are examined in tandem with heterosexuality and homosexuality.

Statistics and the Making of an Average Sexual Appetite

By the mid-1940s, medical and scientific authorities believed that homosexuality resulted from psychological and social factors, and also that it could not be revealed by an examination of "signs" on the body.[20] Kinsey, who in 1938 had been involved in teaching a marriage course with colleagues, was dissatisfied by the lack of unbiased information available to students on sexuality, reproduction and marriage.[21] He also deplored the

[17] Terry, *An American Obsession*, 181.

[18] Ibid., 217.

[19] George William Henry, *Sex Variants: A Study in Homosexual Patterns* (New York: Paul Hoeber & Sons, 1941).

[20] Terry, *An American Obsession*, 297.

[21] For an account of Kinsey's intellectual trajectory to the study of sexuality, see chapters six and seven in Vern L. Bullough, *Science in the Bedroom: A History of Sex Research* (New York: Basic Books, 1994).

lack of large-scale and thorough scientific research on sexuality. *Sexual Behavior in the Human Male* (henceforth *Male*) was released to the public in January 1948 and propelled Kinsey to iconic status in mid-twentieth-century North America. The report quickly became a best seller, generating passionate responses from supporters and detractors who saw the volume as shedding light on North American social and cultural customs as well as providing insights into the country's future.[22] Kinsey became a regular topic of magazine and newspaper articles across the United States. His influence on sexual mores in twentieth-century America was so profound that the historian Vern L. Bullough declared that "sex before Kinsey was radically different than it was after."[23] The *Male* report and the second volume of the Kinsey studies, *Sexual Behavior in the Human Female* (henceforth *Female*), which was published in 1953, "propelled sex into the public eye in a way unlike any previous book or event had done."[24] Trained as a zoologist, Kinsey endeavoured to amass data and disseminate scientific knowledge on sexuality. In fact, for Kinsey, any behaviour, however controversial, could be explained by referring to biological impulses.[25] The team of researchers led by Kinsey compiled thousands of case histories, which they coded, analysed and mapped into statistical representations of patterns of sexual activity.

In the unpublished lecture "Biological Aspects of Some Social Problems" (1935), Kinsey deplored the "ignorance of sexual structure and physiology, of the technique fundamental in the normal course of sexual activities."[26] He largely blamed religious authorities and, as James H. Jones writes, "considered the church the family's worst enemy."[27] Kinsey was deeply committed to outlining and popularising a scientific view of sexuality, one that would be confirmed by solid biological data. By July 1938, he had begun to collect the sexual histories of participants. In 1938, sixty-two histories were collected over six months. Over the years,

[22] Drucker, "Intellectual Responses to *Sexual Behavior in the Human Male*," 79.

[23] Vern L. Bullough, "Sex Will Never Be the Same: The Contributions of Alfred C. Kinsey," *Archives of Sexual Behavior* 33, no. 3 (2004): 277.

[24] D'Emilio and Freedman, *Intimate Matters*, 285.

[25] Irvine also notes that "Kinsey was an essentialist for whom 'natural' equaled good." Irvine, *Disorders of Desire*, 27.

[26] Alfred C. Kinsey "Biological Aspects of Some Social Problems," quoted in James H. Jones, *Alfred C. Kinsey: A Life* (New York: W. W. Norton & Company, 1997), 307.

[27] Jones, *Alfred C. Kinsey*, 307.

while refining the interview techniques and expanding the number of questions, the number of participants grew, reaching 12,214 by 1947. Kinsey himself recorded 57.6%, or 7036, of the histories used in the *Male* volume.[28] As noted in the introduction to the *Male* report, the researchers collected more data than could be handled.[29]

Kinsey's studies "consolidated the status of the empirical survey of sexuality, the kind of survey that became ubiquitous throughout the post-war world."[30] While the method of survey and the collection of statistics was not unprecedented, the thoroughness and accessibility of the data were quite novel to the public, exemplifying the distinctly "modern" character of the reports. As Sarah E. Igo notes, the Kinsey research was "linked to a new way of knowing, one that was 'modern' in its willing confrontation with stark, difficult realities."[31] The volumes were modern not solely because of the sensitive or "taboo" subject matter but also because of the way the data was presented to the public, "in a technical language of social science, and specifically, quantification."[32] The methodology adopted in the reports gave the data the scientific legitimacy that Kinsey actively sought. The numbers were "spare, clear, and direct."[33] The data seemed both far-reaching—given the large number of participants, 5300 men and 5940 women—and appeared valid precisely *because* they were numbers, rather than "subjective" biographical narratives. As Kinsey's co-researchers later reflected, "the [interview] instrument was singularly devoid of items involving interpretation and introspection."[34] The lack of biographical narratives would also thwart psychiatric or psychoanalytical analysis of data, and physicians in those fields would point to the necessity of these details,[35] arguing that sustained in-depth individual therapy was the "only

[28] Alfred C. Kinsey, Wardell B. Pomeroy and Clyde E. Martin, *Sexual Behavior in the Human Male* (Philadelphia and London: W.B. Saunders Company, 1948), 10–11.

[29] Ibid., 11.

[30] Chris Waters, "Sexology," in *Palgrave Advances in the Modern History of Sexuality*, eds. H. G. Cocks and Matt Houlbrook (New York: Palgrave Macmillan), 49.

[31] Sarah E. Igo, *The Averaged American: Surveys, Citizens, and the Making of a Mass Public* (Cambridge, MA: Harvard University Press, 2007), 246.

[32] Ibid.

[33] Ibid., 247.

[34] Paul H. Gebhard and Alan B. Johnson, *The Kinsey Data: Marginal Tabulations of the 1938–1963 Interviews Conducted by the Institute for Sex Research* (Indianapolis: Indiana University Press, 1979), 11.

[35] Theories of psychoanalysis dominated the field of psychiatry in the United States until the late-twentieth century. The psychoanalyst Edmund Bergler was particularly critical of

valuable clinical material."[36] Using a quantitative rather than qualitative methodology, the *Male* and *Female* studies stand as examples of the rise of statistical calculations in the management of populations. The Kinsey reports further propelled the study of sexuality as a tenable scientific area of research, paving the way for sex researchers such as Masters and Johnson.

The collection of biographical details such as family history and sexual activity was a longstanding practice in sex research. The patient case history, which I discussed in chapter two, enabled physicians and researchers to accumulate detailed information about sexual behaviours, sometimes across the patient's lifetime. The patient's narrative and the practice of anthologising case histories were instrumental to the shaping of categories of "normal" and "perverse" sexuality. By the time the *Male* and *Female* reports were published, personal narratives were common tools for analysing "deviance." Research projects on sexuality that preceded Kinsey tended to be characterised by personal accounts of a smaller number of participants, which were accompanied by the researcher's outline of broader societal implications. The researcher often knew participants as patients, or the participants engaged in written correspondence with the researcher and they never met in person. The combination of a large sample in the Kinsey studies, the turn to quantification and statistics and the widespread accessibility of the reports are indications of a significant break from previous sex research.[37]

A seldom observed aspect of the Kinsey reports was the focus on the frequency of different sexual practices. In the Kinsey reports, sexual appetite was not limited to whether someone had low or excessive desire for sexual activities.[38] It was rather embedded in the question of how sexuality can be both measured and accurately—which, for Kinsey, meant scientifically—represented. The statistics and graphs produced in the

Kinsey's findings and interpretation. Bergler notably rebuked Kinsey and the studies for attempting to normalise homosexuality. See Edmund Bergler, *Homosexuality: Disease or Way of Life?* (New York: Hill and Wang, 1956) and "The Myth of a New National Disease: Homosexuality and the Kinsey Report," *The Psychiatric Quarterly* 22, no. 1–4 (1948): 66–88.

[36] Miriam G. Reumann, *American Sexual Character: Sex, Gender, and National Identity in the Kinsey Reports* (Berkeley and Los Angeles: University of California Press, 2005), 27.

[37] Note that the use of quantification and statistics led some commentators to fault the Kinsey's reports for not attending to contexts of intimacy, love and committed partnership. See Drucker, "Intellectual Responses to Alfred Kinsey's *Sexual Behavior in the Human Male*."

[38] However, this was a feature of both volumes. See for example, Kinsey et al., *Male*, 199 and 237.

reports were themselves shaped by questions of frequency. Beyond the necessity of sexual appetite, which is at the heart of the research, appetite can lend itself to measurement and can reveal a truth to human sexuality. Kinsey expressed distrust of labels such as "frigid, sexually under-developed, under-active, excessively active … hypersexual." He noted that they refer to "nothing more than a position on a curve which is *continuous*."[39] At work in this observation is the notion of *degree*. The Kinsey team was thus focused on the question of "how much?" or "how often?" The *Female* volume outlines that "the statistical data [in both reports] have been largely concerned with the incidences and frequencies of sexual activities that led to orgasm."[40] Sexual object choice certainly loomed large in the interview, but the research team was careful to extract averages rather than identity.[41]

Although lengthy personal narratives were not included in the published Kinsey reports, the interview was central to Kinsey's methodology as he was convinced that through this method, as opposed to a questionnaire or a physiological examination, the participant could *accurately* provide a portrait of her or his sexual history. To circumvent exaggerations and deceit, Kinsey included several tactics such as probing deeper into a specific question and cross checking the account of couples with spouses. For example, under "Cross-checks for accuracy," the *Male* volume outlines how the "best protection against cover-up lies in the use of a considerable list of interlocking questions."[42] As Bullough notes, Kinsey was convinced that he could uncover fallacious responses and his "ingenious coding system was designed to detect the most obvious ones."[43] The interviews were coded using a specific method that the volumes did not explain to maintain "the confidence of the record."[44] The coded records were subsequently transferred to "punched cards for statistical analyses."[45] Drucker observes that the "punched-card operation had function and

[39] Kinsey et al., *Male*, 199 (emphasis added).

[40] Alfred C. Kinsey, Wardell B. Pomeroy, Clyde E. Martin and Paul H. Gerbhard. *Sexual Behavior in the Human Female* (Philadelphia and London: W.B. Saunders Company), 510.

[41] "Males," Kinsey wrote, "do not represent two discrete populations, heterosexual and homosexual … Not all things are black not all things white. It is a fundamental of taxonomy that nature rarely deals with discrete categories … The living world is a continuum in act and every one of its aspects." Kinsey et al., *Male*, 639.

[42] Ibid., 54–55.

[43] Bullough, *Science in the Bedroom*, 175.

[44] Kinsey et al., *Male*, 71.

[45] Ibid.

meaning, and all were automated for efficiency and flexibility."[46] The cards were then used to produce, analyse and compile statistics on the frequency of sexual activities. The translation of sexual appetite into statistical data was thus mediated by the use of machines of calculation.

In the nineteenth century, statistical representations of different facets of life, such as birth and death rates, which were mostly collected through coronial institutions, census surveys and stringent reporting mechanisms, came to inform what Michael Power calls a "political arithmetic."[47] Governments turned to arithmetic and calculation to inform policies related to public health. The population, in turn, was represented by numbers and data to render governance feasible. For Foucault, modern governance was defined by "the administration of bodies and the calculated management of life."[48] As Cryle argues, Foucault's phrase, "*la gestion calculatrice de la vie*"[49] can be translated to "the management of life with a calculator," exemplifying how the processing of numbers into statistics gave "full scientific meaning to new forms of governmental practice."[50]

The collection of data by medical institutions and its representation in charts and statistics served to constitute kinds of people, that is to say, it classified individuals as belonging to a certain category of person. The accumulation and processing of numbers on certain aspects of life became a means to achieve a calculated, mathematical understanding of individuals and their practices with impartiality. As Theodore M. Porter notes, "[q]uantification is not merely a strategy for describing the social and natural worlds but a means of reconfiguring them."[51] Importantly for this genealogy of sexual appetite, representing subjects and their sexual practices in the form of numbers produced a relational and social understanding

[46] Donna J. Drucker, "Keying Desire: Alfred Kinsey's Use of Punched-Card Machines for Sex Research," *Journal of the History of Sexuality* 22, no. 1 (2013): 109.

[47] Michael Power, "Counting, Control and Calculation: Reflections on Measuring and Management," *Human Relations* 57, no. 6 (2004): 766. See also, Marc Trabsky, *Law and the Dead: Technology, Relations and Institutions* (Abingdon: Routledge, 2019).

[48] Michel Foucault, *The Will to Knowledge: The History of Sexuality, volume 1*, trans. Robert Hurley (London: Penguin Books, 1978), 140.

[49] Michel Foucault, *Histoire de la sexualité 1: La volonté de savoir* (France: Gallimard, 1976), 184.

[50] Peter Cryle, "The Average and the Normal in Nineteenth-Century French Discourse," *Psychology & Sexuality*, 1, no. 3 (2010): 217.

[51] Theodore M. Porter, "Making Things Quantitative," *Science in Context* 7, no. 3 (1994): 389. See also Theodore M. Porter, *Trust in Numbers: The Pursuit of Objectivity in Science and Public Life* (New Jersey: Princeton University Press, 1995).

of the person. In other words, data enables comparative practices and encourages subjects to judge their lives in relation to the average; an average sexual appetite as reflected in the frequency of certain activities. In addition, Kinsey affirmed that the "high incidences shown for several types of sexual activity are not exaggerations of the fact, but every calculation indicates that they are understatements, if they are in error at all."[52] The Kinsey projects translated sexuality into statistics and hence functioned to produce a calculable subject, an individual whose sexual appetite can be turned into aggregated numerical values.

Kinsey's commitment to the truth-value of numerical data situated techniques and procedures of data collection and management as central to the production of knowledge. He expressed confidence in statistical calculations: "[Statistical analysis] is, precisely, the function of a population analysis to help in the understanding of particular individuals by showing their relation to the remainder of the group."[53] He rejected ideas that "average individuals do not really exist, and that measurements of such hypothetic individuals provide no insight into particular persons with whom the clinician must deal."[54] For Kinsey, such attitudes revealed a misunderstanding of statistical analysis, averages and norms. While not fully rejecting the importance of studying the individual, he posited that locating individual sexual practices within a study of a large group was much more productive. Certain sexual behaviours, the outliers, may be rare or common among the group. However, rarity in this conceptualisation is not a question of abnormality but rather one of statistical incidence within a studied population.[55] This approach is aligned with Kinsey's affirmation that all behaviours have biological underpinnings and are hence not "abnormal." As he wrote: "No individual has a sexual frequency which differs in anything but a slight degree from the frequencies of those placed next on the curve. Such a continuous and widely spread series raises a question as to whether the terms "normal" and "abnormal" belong in a scientific vocabulary."[56] Again, this comment demonstrates Kinsey's concern with matters of degree and frequency in sexuality. The language of

[52] Kinsey et al., *Male*, 121.

[53] Ibid., 20.

[54] Ibid.

[55] Ibid., 21.

[56] Kinsey et al., *Male*, 199. See also Alfred C. Kinsey, Wardell B. Pomeroy, Clyde E. Martin and Paul H. Gerhard, *Concepts of Normality and Abnormality in Sexual Behavior* (New York: Grune & Stratton, 1949).

appetite thus *flows through* his research. Kinsey's understanding of sexuality deployed in the *Male* and *Female* volumes and his methodology are imbued in sexual appetite. The frequency of practices was revealed by techniques of measurement.

In *Inventing Our Selves*, Rose argues that "truth becomes effective to the extent that it is embodied in technique."[57] The information on sexual practices derived from data were viewed as "uncontaminated," or at least, uninfluenced by what Kinsey considered "speculation and armchair theorizing."[58] Kinsey wanted the studies to be "soundly buttressed by quantified data."[59] Thus, the Kinsey reports represented statistical knowledge as a matter of *technique*, "rooted in attempts to organize experience according to certain values."[60] The "truth" revealed in quantifications of sexual appetite and the data produced from quantifications was also technical.

Kinsey might have seen his task as simply reporting numbers without subjective interpretation, but statistics, Porter remarks, "participate actively in the formation of individual and collective identities."[61] And the authority of data depends on external validation. In other words, physicians, statisticians, as well as non-experts, continuously participate in a network of interpretation and evaluation. Hence, equipment such as the punched-card machines used to code data from interviews, and the statistics and charts presented in the reports, impact on the social and intimate lifeworlds of individuals. As Rose writes, "Vocabularies of calculation and accumulations of information go hand in hand with attempts to invent techniques by which the outcomes of calculative practice … can be translated into action upon the objects of calculation."[62] While the punch-card machine could process large quantities of data rapidly as well as create anonymity by "aggregating sex history data on card,"[63] the machine also located human properties in a network of administrative and scientific procedures.

[57] Nikolas Rose, *Inventing Our Selves: Psychology, Power, Personhood* (Cambridge: Cambridge University Press, 1998), 89.

[58] Gebhard and Johnson, *The Kinsey Data*, 11.

[59] Ibid.

[60] Rose, *Inventing Our Selves*, 89.

[61] Porter, "Making Things Quantitative," 400.

[62] Nikolas Rose, "Calculable Minds and Manageable Individuals," *History of the Human Sciences* 1, no. 2 (1988): 185.

[63] Drucker, "Keying Desire," 113.

Kinsey strongly believed that sexuality needed to be studied mathematically in order to avoid moral judgements. The introduction to the *Male* report affirmed that the research is on what people do, and not "what they should do, or what kinds of people do it."[64] The introduction announced the study as "an unfettered investigation of all types of sexual activity, as found among all kinds of males."[65] The studies then attempted to avoid producing "kinds of people" as abnormal or perverse: "Kinsey viewed sexual identity categories as a way for society to inflict harm on individuals for acting on desires outside social norms."[66] Kinsey distanced his studies from definitions of sexual identities, such as homosexuality. Rather than defining homosexuality in terms of a set of practices, self-identification, experience and behaviour, or seeking the traces of homosexuality on bodies, Kinsey turned to the concept of "sexual outlet." The measure of "sexual outlet" referred to how (and how often) the participants achieve orgasm. Hence, sexual appetite emerged here in terms of its (possible) outcome—the orgasm—and how *often* an individual participates in certain activities to reach orgasm.

In a study of the production of knowledge on research objects, Hacking argues that the making up of "kinds of people" involves five components: classification, people or research subjects, institutions, knowledge and experts.[67] These parts continually influence each other and work to produce what and how practices or circumstances are known. Hacking further suggests that knowledge and discovery depend on counting, quantifying and the production of norms.[68] When social and intimate aspects of life are coded into mathematical data and released to the public, notions of norms do not altogether disappear, as the reactions to the Kinsey reports demonstrate. Thus, while the measure of "sexual outlet," and the interconnected sexual appetite, circumvent identity and justifications, the statistics, nonetheless, produce a sketch of who has "average" sex, what constitutes "average" sexuality and provide an opportunity for individuals to measure themselves against the average.

[64] Kinsey et al., *Male*, 7.

[65] Ibid.

[66] Donna J. Drucker, "Male Sexuality and Alfred Kinsey's 0–6 Scale: Toward 'A Sound Understanding of the Realities of Sex'," *Journal of Homosexuality* 57, no. 9 (2010): 1106.

[67] Ian Hacking, "Kinds of People: Moving Targets," *Proceedings of the British Academy* 151 (2007): 285–318.

[68] Ibid., 305–309.

The statistical representation of sexual patterns contributed in the 1950s to comparative exercises where individuals could assess their own practices against the data represented. By using the "all-purpose measure of outlet, the scientist's numerical charts invited measurements against the mean."[69] Indeed, as Igo notes, in a study on surveys and American "mass society," "there is much evidence to suggest that individuals *were* using Kinsey's data as … [a] standard by which to classify their own behaviour."[70] Statistical models unveiled a distributed variance throughout the population.[71] One of the pivotal breakthroughs of the *Male* volume, the 0–6 scale—0 being "exclusively heterosexual with no homosexual" and 6 "exclusively homosexual"[72]—was an attempt to capture the fluidity of male sexuality and their object choices. The plotting of behaviours on a continuum of heterosexual to homosexual demonstrated not only that homosexuality was more common than people had imagined, but also that sexual object choice could change across one's lifetime. For Katz, the so-called Kinsey scale suggests that "there are degrees of heterosexual and homosexual behaviour and emotion. But that famous continuum also emphatically reaffirmed the idea of a sexuality divided between the hetero and homo."[73] However, the publication of the two Kinsey studies, while certainly revealing broad variations in sexual activity and situating individuals on a continuum, also produced an "average" subject with an "average" sexual appetite. The reports held on to the question of sexual appetite and mobilised it throughout their interpretation of data.

Comparative exercises allowed individuals to estimate how common or normal their sexual experiences were. While the Kinsey team was careful to distance the studies from the question of what normal sexuality is, the statistics offered an evaluative terrain. For example, to the question "Do you notice any particular time of the month when it is easier for you to get sexually excited?" 24.9% of college-educated women replied no, while 34% replied that they noted a difference just before menstruation.[74] This question invited a reflection, a look to one's bodily experiences to be attuned to what was happening periodically to the body. Of course, the questions were answered on the day of the interview, hence the participant

[69] Igo, *The Averaged American*, 261.
[70] Ibid., 264 (emphasis original).
[71] Ibid., 303–304.
[72] Kinsey et al., *Male*, 638.
[73] Jonathan Ned Katz, *The Invention of Heterosexuality* (New York: Dutton, 1995), 97.
[74] Gebhard and Johnson, *The Kinsey Data*, 146.

did not have time to reflect or diarise their experience beforehand. Nonetheless, it is notable that participants would be encouraged to evaluate their incidence of sexual arousal. As a result, readers of the reports may have engaged in such reflections and paid attention to their bodies. This attention to time and periodicity is quite similar to Davis' and Hamilton's research, discussed earlier in this chapter. Frequency and duration were recurrent features of the interviews which, I argue, demonstrate how sexual appetite and dimensions of quantity and rate of occurrence operate in the reports. Participants were, for instance, asked how often they engaged in sex in the missionary position with the man on top, how often with the woman on top, or how often they had intercourse while sitting, standing and "with the female's back towards the male."[75] The interviews covered incidence, frequency, duration and acts, rather than identities, creating averages of intensity, quality, quantity and behaviour. Interview data, while inviting participants to reflect and report was, however, distanced from the patient case history that had so characterised the field of sexual science.

Observation, Norms and Appetite in the Laboratory

In the opening pages of *Human Sexual Response*, Masters and Johnson praised the Kinsey team for opening an avenue of research that had been severely restricted as a legitimate object of study for a long time. They called the Kinsey study a landmark of "sociologic investigation" but considered that a crucial limitation was the lack of attention to the biological basis of human sexual response.[76] Two questions, they asserted, remained to be answered: "What physical reactions develop as the human male and female respond to effective sexual stimulation?" and "Why do men and women behave as they do when responding to effective sexual stimulation?"[77] While indebted to Kinsey, Masters and Johnson privately considered the *Male* and *Female* studies to be flawed because they relied on recollections rather than direct observation.[78] Despite the lasting influence that Masters and Johnson had on sex research, the turn to physiology and the responses of genitalia was not completely new. Physicians such as

[75] Ibid., 302.

[76] William H. Masters and Virginia E. Johnson, *Human Sexual Response* (Boston: Little, Brown and Company, 1966), 3.

[77] Ibid.

[78] Thomas Maier, *Masters of Sex: The Life and Times of William Masters and Virginia Johnson, the Couple Who Taught America How to Love* (New York: Basic Books, 2009), 97.

Van de Velde and Félix Roubaud had already conducted empirical studies of physiology, while in the United States, the work of CSSV, discussed above, and particularly the empirical research and inventions of the obstetrician Robert Latou Dickinson, were especially important.[79] Dickinson conducted studies in the first half of the century, and also created a "phallus-shaped glass tube" through which he could study the responses of the vagina during orgasm.[80] It is worth noting that in *The Single Woman* (1934), Dickinson and Lura Beam studied patients at different points in time. This approach enabled them to identify sexual appetite as a fluctuating feature of the life course: they illustrated twenty cases of how "'passion and frigidity' could appear and disappear."[81] Like many of their contemporaries and predecessors, Dickinson and Beam's study has been examined by scholars chiefly in terms of its approach to homosexuality and women more broadly.

However, they constitute an important contribution in the genealogy of sexual appetite. Indeed, in their study, sexual appetite operates through an attention to arousal, frequency and intensity of desire.[82]

Irvine identifies the timing of Masters' and Johnson's publications as key to understanding their rise to prominence. While their predecessors conducted research in secret and results rarely made their way to the public, the social mores of the late 1960s, alongside the exposure of Kinsey's reports, meant that Masters' and Johnson's research was better received. The 1960s in the United States was a decade of social and political change. A combination of activist mobilisations, especially antiwar and civil rights activism, second-wave feminism and gay liberation movements, alongside the end of two World Wars, and inventions such as the birth control pill, resulted in social and sexual transformations. Much like Kinsey and in line with the spirit of social liberalism, Masters and Johnson did not associate

[79] See for example, Van de Velde, *Ideal Marriage*, and Félix Roubaud, *Traité de l'impuissance et de la stérilité chez l'homme et la femme comprenant l'exposition des moyens recommandés pour y remédier* (Paris: J B Ballière, 1855).

[80] Irvine, *Disorders of Desire*, 54.

[81] Bullough, *Science in the Bedroom*, 110. See Robert Latou Dickinson and Lura Beam, *The Single Woman: A Medical Study in Sex Education* (Philadelphia: The Williams & Wilkins Company, 1934), 144.

[82] For example, George Chauncey Jr, "From Sexual Inversion to Homosexuality: The Changing Medical Conceptualization of Female 'Deviance', " in *Passion and Power: Sexuality in History*, eds. Kathy Peiss and Christina Simmons (Philadelphia: Temple University Press, 1989), 87–117, Faderman, *Odd Girls and Twilight Lovers*, and Terry, *An American Obsession*.

female sexuality with motherhood, but instead "viewed women as individuals with desires upon which they acted"[83] and emphasised the importance of listening to women's descriptions of their sexual desires.[84] In *Human Sexual Response*, they deplored "[d]ecades of 'phallic fallacies'" in research on the clitoris and argued that current understandings of the clitoral function were flawed because research had been "uninformed by female subjective expression."[85] As several scholars have noted, while Masters' and Johnson's works were welcomed by feminists as "liberating," for instance, their challenge to ideas on the vaginal orgasm, the sexologists "organized their data around their own conservative interpretations."[86] They elided questions of gender inequality and reproduced normative cultural ideas on relationships and marriage. In their second book, *Human Sexual Inadequacy*, published in 1970, Masters and Johnson noted that sex therapy and the learning of proper technique could in fact strengthen and save marriages, and render their book "obsolete in the next decade."[87]

Human Sexual Response was an instant success and sold out within three days of its publication. The public was eager for "unbiased" information on sexual activity, and the authors' "self-presentation as solid, no-nonsense scientists"[88] contributed to the popularisation of their research as objective. Masters and Johnson introduced the human sexual response cycle (HSRC) in *Human Sexual Response*. The cycle comprised excitement, plateau, orgasm and resolution; the phases were established as occurring in women and men.[89] While Kinsey sought to catalogue the variations of human sexuality, Masters and Johnson established diversity as a variation between normal and abnormal, and continually privileged married heterosexual monogamy. Their works reveal a focus of sexual medicine on physiology and the attention to the "mechanics" of sexual appetite.

[83] Jane Gerhard, *Desiring Revolution: Second-Wave Feminism and the Rewriting of American Sexual Thought, 1920 to 1982* (New York: Columbia University Press, 2001), 52.

[84] Irvine, *Disorders of Desire*, 65.

[85] Masters and Johnson, *Human Sexual Response*, 45.

[86] Irvine, *Disorders of Desire*, 65. This is further apparent in their later works, see William H. Masters and Virginia E. Johnson, *The Pleasure Bond: A New Look at Sexuality and Commitment* (Boston: Little, Brown and Company, 1974), *Homosexuality in Perspective* (Boston: Little, Brown and Company, 1979) and *Heterosexuality* (New York: Harper Collins, 1994).

[87] William H. Masters and Virginia E. Johnson, *Human Sexual Inadequacy* (Boston: Little, Brown and Company, 1970), v.

[88] Irvine, *Disorders of Desire*, 66.

[89] Masters and Johnson, *Human Sexual Response*, 4–5.

Rather than letting the subject speak and provide an account of the frequency of sexual activity, as Kinsey had done, Masters and Johnson looked to the body's functions to unveil biological truths at the heart of how sexual appetite, and sexuality more broadly, works. The behavioural interpretation asserts that "sexual responses are natural 'unconditioned' reactions and dysfunctional symptoms are learned inhibitions."[90] Consequently, Masters and Johnson recommended that by affirming the nature of cyclic sexual responses, individuals would be able to attain this ideal of consistent, lasting and innate appetite.

Human Sexual Inadequacy focused on problems of functioning, to which sexual appetite was particularly significant. This work also devoted around ninety pages to principles of sex therapy, which included instructions on how to record the patient's medical history, alongside goals to be attained each day. Hence, the subject was not fully silenced. Instead, a combination of observation and measurement emerged in the works of Masters and Johnson. While marriage therapy had already existed for several decades, the industry of sex therapy was still quite recent and expanded sharply after the publications of *Human Sexual Response* and *Human Sexual Inadequacy*. In addition to machines for measurement, the role of observation—the ability to peruse bodies and document their reactions to stimuli—was central to these works.

To study the manifestations and flows of sexual appetite inside their laboratory in the late 1950s, Masters and Johnson selected 312 men and 382 women who were observed during masturbation and sexual intercourse. The participants, predominantly white, educated and upper middle class, provided the data on which the HSRC is based.[91] Data was gathered from machines during each phase of the HSRC—excitement, plateau, orgasm and resolution. The participants were observed during a range of activities including masturbation with and without a vibrator, sexual intercourse with one partner in the supine position, coitus with "Ulysses"[92]—the transparent camera-equipped plastic phallus—and female breast stimulation. The HSRC and the associated scenarios involved coupled heterosexual[93] or solo acts only. *Human Sexual Response* also identified

[90] Anna Leeming and Paul Brown, "An Eclectic or Integrative Approach to Sex Therapy?" *Sexual and Marital Therapy* 7, no. 3 (1992): 285.

[91] Masters and Johnson, *Human Sexual Response*, 11–12.

[92] For an account of Ulysses in action, see Maier, *Masters of Sex*, 100.

[93] They note: "The sensitivity of the rectum to stimulation was adjudged essentially equal between the two sexes by gross clinical observation. It must be remembered, however, that

key differences in the male and female anatomy—they identify the clitoris, for example, as a "unique organ"[94] of sexual appetite—while affirming essential similarities in male and female HSRC.

The research of Masters and Johnson was underpinned by a range of normative assumptions about sexuality, especially with regards to marriage. They continually elevated the institution of marriage, focusing their experiments on "family units" or "marital partners," terms they used in *Human Sexual Response* and *Human Sexual Inadequacy*. Information on sexual appetite, and on the female orgasm in particular, was identified as a way to strengthen the marital unit:

> With orgasmic physiology established, the human female now has an undeniable opportunity to develop realistically her own sexual response levels. Disseminating this information enables the male partner to contribute to this development in support of an effective sexual relationship within the marital unit.[95]

The orgasm was identified as the "ultimate point in progression" of the HSRC.[96] It was represented as the necessary outcome of sexual activity, and it functioned as an indicator of "normal" sexual function. Indeed, participants unable to orgasm through intercourse or masturbation were not included in their research.[97] The establishment of orgasm as the "goal" of sexual activity, as well as the manifestation of sexual satisfaction, enabled Masters and Johnson to subsequently offer mechanical solutions to any failures of sexual appetite *en route* to orgasm, which were prominently featured in their next publication, *Human Sexual Inadequacy*.

The HSRC mobilised appetite as a prerequisite to sexual activity. "Excitement," the first stage of the cycle, "develops from any source of somatogenic or psychogenic stimulation. The stimulative factor is of major import in establishing increment of sexual tension to extend the cycle."[98]

material of homosexual content has not been included in this review." Masters and Johnson, *Human Sexual Response*, 200.

[94] "The clitoris is a unique organ in the total human anatomy. Its express purpose is to serve both as receptor and transformer of sensual stimuli… No such organ exists within the anatomic structure of the human male." Masters and Johnson, *Human Sexual Response*, 45.

[95] Ibid., 138.

[96] Ibid., 127.

[97] Ibid., 311.

[98] Ibid., 5.

It is significant that Masters and Johnson omitted references to terms that had been circulating in 1960s sexology, and would continue to influence the field: drive, desire, libido and so on. In seeking to develop a scientific understanding of the physiology of sex, they avoided explaining exactly what they meant by "excitement" and instead focused their study on charting frequency and bodily responses using laboratory equipment. I would suggest that the term "excitement" captures the ebbs and flows of sexual appetite, which continue to circulate in the rest of the HSRC. For Leonore Tiefer, the omission of "drive from their model eliminated an element which is notoriously variable within populations and paved the way [for] a universal model seemingly without variability."[99] While excluding drive from their work, sexual appetite is present in the researchers' attention to questions of rhythm, performance, frequency and duration in sexual activity.

Masters and Johnson assert in the final chapter of *Human Sexual Response* that norms of human sexuality do not exist, hence the difficulty of determining how common are the responses of their participants. Readings from machines are essential. In this chapter, they detail the sexual history of four selected participants, including the frequency of sexual intercourse, masturbation and any loss of sexual appetite. Their histories are presented as representative of the participants of the study. The chapter closes with an affirmation that any sexual inadequacy encountered throughout the course of their research could be easily explained by situational factors (e.g. performing while being watched by clinicians) and easily rectified. Orgasmic difficulties during intercourse are identified as the major issue for women.[100] Overall, they claim, the constant feature of all subjects was a "basic interest in and desire for effectiveness of sexual performance."[101] This basic existence of sexual appetite is again reiterated in the opening passages of *Human Sexual Inadequacy*.[102] In the work of Masters and Johnson, sexual appetite is mobilised at the centre of the HSRC; it is essential to sexual activity. It also emerges as a feature that can be assessed with a degree of precision and it manifests itself in different

[99] Leonore Tiefer, "Historical, Scientific, Clinical and Feminist Criticisms of 'The Human Sexual Response Cycle' Model," *Annual Review of Sex Research* 2, no. 1 (1991): 4.

[100] Masters and Johnson, *Human Sexual Response*, 313.

[101] Ibid., 315.

[102] Masters and Johnson, *Human Sexual Inadequacy*, 12.

ways depending on what is being assessed and whether a machine or the observer is used.

Participants in the Masters and Johnson research project provided the raw data, which was supplemented by figures derived from techniques of observation and measurement. A cartography of sexual appetite emerged from the data accumulated. In the opening pages of *Human Sexual Response*, Masters and Johnson noted:

> The techniques of defining or describing the gross physical changes which develop during the human male's and the female's sexual response cycles have been primarily those of direct observation and physical measurement ... regardless of the observer's training and considered objectivity, reliability of reporting has been supported by many of the accepted techniques of physiologic measurement and the frequent use of color cinematographic recording in all phases of the sexual response cycle.[103]

Their explanation of methodology revealed a reliance on optics of sexual activity and any human error—whether from the observer or the research participant—was addressed by the objectivity afforded by machines and cinematography.

The documentation of the human sexual response was mediated through machines of measurement and clinical observation. Importantly, the machines were not solely measuring responses in genitals; they were also accumulating data on heart rate, blood pressure, perspiration, breast engorgement, the erection of nipples and the so-called sex flush,[104] which could be observed on the surface of bodies, especially on the breasts of women and on the epigastrium of some men. Television screens recorded the reactions of bodies in various stages of sexual activity. Sexual appetite itself was being recorded, its manifestations on the skin and inside the body, and translated by machines. By asserting that male and female responses to sexual stimuli were essentially similar, heterosexual couples could be "taught" how to better stimulate and respond to their partners. The laboratory of sex was hence also a pedagogical space where sexual technique could be instructed and sexual appetite could be optimised. The researchers accumulated notes about their participants, charting changes

[103] Masters and Johnson, *Human Sexual Response*, 4.
[104] Ibid., 31.

over time and in different configurations, for example, sexual performance alone and with a partner.

In their observations, Masters, Johnson and their team of researchers paid attention to how bodies moved, what positions were preferred and what movements enhanced orgasms. The penis-camera—or the "artificial coital equipment"[105]—allowed for the collection of data pertaining to the interior of women's vaginas. How the vagina reacted as an interior and penetrable space during sexual activity was positioned as crucial to sexual function. The equipment was adjustable to suit varying physical needs depending on the female subject. The comfort and technique of the research participant was carefully documented in terms of "rate and depth of penile thrust."[106] All the machines used for women were used for the study of men's sexual responses, except for the penis-camera.[107] While Masters' and Johnson's work argued that women who experienced clitoral orgasms, as opposed to vaginal ones, were normal and not frigid, they nonetheless relied on a number of heteronormative biases during the course of their research.[108] For instance, they affirmed that the clitoris was stimulated "every time the female responds to a male thrust" and that there was no "physiological difference among clitoral orgasm, vaginal orgasm, breast orgasm or … orgasm through fantasy."[109] The research laboratory, in Masters' and Johnson's terms at least, classifies the body's openings as legitimate spaces for sexual activity and maps the diffusions and absences of sexual appetite: "Those machines thus created and delineated two new types of sexual space: the cinematically mapped interior of

[105] Ibid., 21.

[106] Ibid.

[107] Donna J. Drucker, *The Machines of Sex Research: Technology and the Politics of Identity, 1945–1985* (Dordrecht: Springer, 2014), 58.

[108] When asked about future research into homosexuality, Masters replied that they hoped "to move into some concept of sexual reversal for those who wish it." Playboy Magazine, "Playboy Interview: Masters and Johnson," *Playboy Magazine* 15, no. 5 (May 1968): 202.

[109] Playboy, "Playboy interview: Masters and Johnson," 80. See also Masters and Johnson, *Human Sexual Response*, 58. The claims on penile penetration and female orgasm were criticised by several researchers. See Shere Hite, *The Hite Report: A National Study of Female Sexuality* (New York: Seven Stories Press, 1976), Alix Shulman, "Organs and Orgasms," in *Women in Sexist Society: Studies in Power and Powerlessness*, eds. Vivian Gornick and Barbara K. Moran (New York: Signet Books, 1972), 296, and Elisabeth Anne Lloyd, *The Case of the Female Orgasm: Bias in the Science of Evolution* (Cambridge, MA: Harvard University Press, 2005).

women's bodies and heterosexual bedrooms in which men and women now had similar scientifically proven capacities for sexual pleasure."[110]

The gaze of the clinician, impartial, poised and investigative, enhanced the creation of a space for measuring sexual appetite: the laboratory of sex. The research methodology was active in "implanting a technology of observation directly into the body studied—a technique that joins technology and the living body."[111] In the course of the Masters and Johnson experiments, excitement and the development of the sexual response emerged as scientific certitudes that could be observed (and surveyed) beyond the physiological changes occurring in genitals: the research method involved observing changes and reactions inside the body and mapping them on the surface of the body. As Jagose remarks, the "notion that certain medical imaging technologies speak the body's truth is underwritten by the related notion that what the body speaks *is* truth."[112] The reactions and transformations that the body exhibits are interpreted as authentic; observation and measurement merely *expose* these to view, rather than condition the body's manifestations.

In *The Birth of the Clinic*, Foucault describes transformations in the role of vision in Western medical thought. The clinical gaze was able to read the patient's body and excavate knowledge of the human body as bound by processes and patterns pertaining to functionality. Foucault notes that the clinical gaze did more than just observe the body; physical signs were to be documented and mapped. As he writes, "in clinical medicine, *to be seen* and *to be spoken* immediately communicate in the manifest truth of disease of which it is precisely the whole *being*. There is disease only in the element of the visible and therefore statable."[113] The ability of the clinical gaze to capture the totality of function and disease became central to medicine in the nineteenth century. Charcot's use of photographic techniques to develop knowledge on hysteria at *La Salpêtrière* exemplified the importance of the observer to designate specific moments in the development of symptoms. The photographic representation of hysteria, as well as the ability to point to specific moments in the disease and document the pro-

[110] Drucker, *The Machines of Sex Research*, 46.

[111] Lisa Cartwright, *Screening the Body: Tracing Medicine's Visual Culture* (Minnesota: University of Minnesota Press, 1995), 24.

[112] Annemarie Jagose, *Orgasmology* (Durham and London: Duke University Press, 2013), 170 (emphasis original).

[113] Michel Foucault, *The Birth of the Clinic: An Archaeology of Medical Perception*, trans. A. M. Sheridan Smith (New York: Vintage Books, 1994 [1973]), 95 (emphasis original).

gression of symptoms, contributed to the establishment of the clinic as an institution. In Masters' and Johnson's research laboratory, observation was also pivotal to their method. The researchers augmented the clinical gaze and emphasised its importance to the process of constructing norms of sexual appetite.

With ambition to uncover "unbiased" insights into human sexuality, the use of imaging technologies in the research laboratory of Masters and Johnson reflects a form of "mechanical objectivity."[114] That is to say, an objectivity characterised by medical and research procedures punctuated by the production of images and data. Objectivity is "blind sight, seeing without interference, interpretation, or intelligence."[115] The aspiration of objectivity is the access to the body's truisms without the interference of human subjectivity. While the observer remains present, the use of technological equipment fosters the observation of "an encoded inscription of an activity,"[116] rather than a subjective description of what is occurring. In the case of Masters' and Johnson's research, we find the representation of sexual appetite in images and graphs: diagrams charting changes in various areas of the body.[117] Through the observation and measurement of appetite, the body comes to be understood as an apparatus, whose parts function in specific ways to achieve a satisfying *dénouement*. The emphasis on the similarities between the female and male sexual response further strengthens functionality and technique in sexual activity.

Masters and Johnson used several machines to collect information on the sexual response of their participants including "a film camera, a penis-camera ... tape recorders, a colposcope, an electroencephalogram, an electrocardiogram, a respirator, and a pH meter with electrode assembly."[118] The range of instruments enabled the researchers to produce an understanding of sexual appetite, and human sexuality more broadly, that was supported by the accumulation of data and images. In *Screening the Body*, Lisa Cartwright examines the use of motion picture apparatus in the study of physiology. She identifies the role of film as a reflection of "the emergence of a new set of optical techniques for social regulation. In laboratory

[114] Jagose, *Orgasmology*, 169–174.

[115] Lorraine Datson and Peter Gallison, *Objectivity* (New York: Zone Books, 2007), 17.

[116] Cartwright, *Screening the Body*, 27.

[117] The male and female sexual response cycles appear on page 5 of *Human Sexual Response*. See also pages 35 and 175 for electrocardiograms and a graph of the "vaginal environment and seminal-fluid content" on page 93.

[118] Drucker, *The Machines of Sex Research*, 47.

culture, medical practice, and beyond, we see the emergence of a distinctly suveillant [sic.] cinema."[119] Surveillance, Foucault and others have argued, is integral to the production of norms. It is a method of gathering and disseminating knowledge ***and*** disciplining bodies. In the research laboratory of Masters and Johnson, the machines that were involved in mapping the sexual responses of the body on an intimate and interior level contributed to the development and consolidation of a norm of sexual appetite.

Masters' and Johnson's research laboratory produced norms of sexual appetite, while mapping the body's erotic potentiality. Drucker explains that "[t]he concept of sexual space concerns the complex relationships between persons, objects, and behaviors and the ways that those relationships change over time."[120] Sexual space brings forth an enactment of sexual activity mediated by objects as well as human partners. The processing of data and the mapping of bodily changes and reactions produced knowledge on sexual appetite. At the same time, Masters and Johnson derived a set of behaviours as constituting "normal" body responses. Norms, as Canguilhem, Foucault and others remark, invoke a power relation with institutions, cultural and scientific beliefs and indeed knowledge itself: "The normal is not a static or peaceful, but a dynamic and polemical concept."[121]

The clinical gaze, Foucault writes, must be able to draw conclusions as soon as it "perceives a spectacle."[122] In Masters' and Johnson's clinic, the observers intently scrutinised and gathered data on sexual appetite, performance and shortcomings to promote technique and aptitude. The spectacle that was laid bare in the space of the laboratory was observed for what it could reveal about the mobilisation of sexual appetite and optimisation of sexual pleasure. As Jagose argues, the ability to represent and document the body during orgasm relies on the erasure of the orgasmic subject into sets of data.[123] In other words, observation requires the removal of individuals from the sexual space—observation is dispassionate and mechanical. The emphasis on detachment, neutrality and objectivity

[119] Cartwright, *Screening the Body*, xiii.

[120] Drucker, *The Machines of Sex Research*, 47.

[121] Georges Canguilhem, *The Normal and the Pathological*, trans. Carolyn R Fawcett (New York: Zone Books, 1991 [1943/1966]), 239. See also Cryle and Stephens, *Normality*.

[122] Foucault, *The Birth of the Clinic*, 107–108.

[123] Jagose, *Orgasmology*, 174.

in the observation of sexual appetite and its paroxysm, orgasm, serves to inscribe those events into a "representational order" of clinical expertise.[124]

The equipment used by researchers brought to light the belief in the ability of medical imaging technology to reveal physiological responses and unveil the truth of bodily responses. Such an approach to the body continues to prevail in contemporary medicine and was not completely foreign to the Kinsey team, which also relied on techniques of observation to understand sexuality. While lengthy interviews recording details of participants' biographies were key to the *Male* and *Female* reports, Kinsey also believed that the observation of sexual activity was important. In the *Female* report, he detailed how viewing human sexual activity is "limited by the custom."[125] Thus, the researchers restricted their study to observing "mammalian sexual behavior," hoping that this would provide insight into human sexuality. Those observations were supplemented with "moving picture records" of sexual activities of fourteen species of mammals: "With the photographic record, it is possible to examine and re-examine the identical performance any number of times and, if necessary, examine and measure the details on any single frame of the film."[126]

These records were also supplemented by accounts provided by researchers who had the opportunity to observe adult humans in a sexual scenario and by parents on the sexual activities of young children. From the data, three chapters in the *Female* report are devoted to the physiology of sexual response and orgasm and the neural mechanism of sexual response. The norm is constructed through the observation of animal, adult human and young children in sexual activities.

In *The Normal and the Pathological*, Canguilhem addresses the question of differentiating between the average and the norm. Canguilhem examines the uses and definitions of normal and average in nineteenth-century medicine and argues that statistical calculations "offer no means for deciding whether a divergence is normal or abnormal."[127] Statistics represent incidences and offer an average "kind," but do not in themselves clarify what constitutes the norm. Thus, Canguilhem argues, it is important to distinguish norm and average as different concepts.[128] Nonetheless, an

[124] Ibid., 174.
[125] Kinsey et al., *Female*, 91.
[126] Ibid.
[127] Canguilhem, *The Normal and the Pathological*, 155.
[128] Ibid., 177.

examination of the Kinsey reports, with their reliance on large numbers and averages, reveal how statistics are perceived as disclosing what constitutes normality in sexual appetite and broader understandings of sexuality. The greater the incidence, the "more normal" a behaviour. However, Canguilhem asserts:

> to consider the average values of human physiological constants as the expression of vital collective norms would only amount to saying that the human race, in inventing kinds of life, invents physiological behaviors at the same time. But are the kinds of life not imposed? ... from the moment several collective norms of life are possible in a given milieu, the one adopted, whose antiquity makes it seem natural is ... the one chosen.[129]

In other words, the norm is always relational to other aspects of the epoch in which it emerges, and Canguilhem himself reminds us that in French, the verb *normer* means to "normalize, to impose a requirement on an existence."[130]

The concept of *norme* emerges in language, referring to normal states and balance, around the late nineteenth century, as symbolised by its inclusion in dictionaries in France.[131] It is worth noting that *norme* first appears in the eleventh volume of the *Grand dictionnaire universel du XIXe siècle* in France in its 1866–1877 edition. The note next to *norme* remarks on the novelty of the word, and cautions the interlocutor to hear it being spoken, and to make a special effort to comprehend it. The dictionary explains: *La norme est l'état régulier, normal, produit par la balance des forces.*[132] This translates as: the norm is the regular state, the normal state, produced by balance. The dictionary goes on to explain that the norm is imposed by society; but, in relation to morality, philosophy and spirituality, it is something to which individuals aspire: *Tout individu*

[129] Ibid., 175.

[130] Ibid., 239.

[131] The word "normal" was used to refer to schools devoted to the training of teachers, *les écoles normales.* The term appeared in the *Dictionnaire de l'académie française* in 1832. See Caroline Warman, "From Pre-normal to Abnormal: The Emergence of a Concept in Late Eighteenth-Century France," *Psychology & Sexuality* 1, no. 3 (2010): 200–213 and, Peter Cryle and Lisa Downing, "Introduction: The Natural and the Normal in the History of Sexuality," *Psychology & Sexuality* 1, no. 3 (2010): 191–199.

[132] Pierre Larousse, *Grand dictionnaire universel du XIXe siècle, (1866–1877), Tome onzième* (Paris: Administration du Grand Dictionnaire Universel, 1874), 1096.

a le droit et le devoir de realiser sa norme[133]—Everyone has the right and the duty to achieve his or her norm. This brief etymological overview of the word "norm" highlights how the normal state also refers to an ideal. As Elizabeth Stephens observes, what is considered normal is "ontologically relational, describing not a fixed thing but an orientation of one thing in relation to another."[134] To be normal, to achieve "normality" in sexual appetite, entails *balancing* different forces while *aspiring* to achieve an ideal of normality. This idea circulates through several techniques studied in this book. Balance indeed continually features as a desirable state in knowledge on human sexuality.

The works of Masters and Johnson reveal the aspiration to normality in sexuality. By scrutinising the body in the setting of the laboratory, Masters and Johnson aimed to remove bodies from their cultural forces. Watching and measuring the body in action, the researchers purportedly extract social norms from sexuality and unveil the *normal* state of the body and its appetites free from the contamination of cultural beliefs. However, as Canguilhem has concisely demonstrated in his work, what is normal is relational to the environment.[135] In his analysis of prominent nineteenth-century French physiologist Claude Bernard's work, Canguilhem highlights Bernard's belief that "the normal is defined as an ideal type in determined experimental conditions rather than as arithmetical average or statistical frequency."[136] As noted in the previous pages, the ideal at the centre of Masters' and Johnson's studies was the orgasm, the pinnacle of sexual appetite. The orgasm was established as a manifestation of sexual satisfaction or the proof that a successful and pleasurable sexual activity has occurred.

While Masters' and Johnson's research demonstrates both their belief in the body's ability to reveal its own norms, and the researchers' favouring of married heterosexual monogamy, Kinsey's methodology was guided by his suspicion towards the interference of cultural and religious norms in knowledge of sexuality. A researcher sceptical of socially sanctioned norms, Kinsey considered his work to reveal averages and incidences but not necessarily to condone any behaviour. Nonetheless, an idea of the

[133] Ibid.

[134] Elizabeth Stephens, "Normal," *TSQ: Transgender Studies Quarterly* 1, no. 1–2 (2014): 143. See also Cryle and Stephens, *Normality*.

[135] Canguilhem, *The Normal and the Pathological*, 143.

[136] Ibid., 152.

normal as a "focal point of discussion, debate, and often disagreement" emerged in the Kinsey reports.[137] The norm, Robyn Wiegman and Elizabeth A. Wilson argue,

> is a dispersed calculation (an average) that enquires into every corner of the world. That is, the measurements, comparisons … that generate the average man do so not in relation to a compulsory, uniform stand, but through an expansive relationality among and within individuals across and within groups[138]

An average then involves all members of a studied population, and "outliers" are always in relation to the centre. In fact, according to Janet Halley, while it can be argued that normalisation "implicitly confirms that the average is good," it is also the case that "normalcy can be articulated as such only if it has outliers." The average, in other words, is dependent upon deviation.[139]

However, as Peter Cryle and Elizabeth Stephens' history of the conceptual slipperiness of the term "norm" demonstrates, norms and averages have a constitutive relationship.[140] At different turns in Western history, the two terms were connected in different ways. In the late nineteenth century in the field of mathematics, Francis Galton, an English statistician, considered the normal to be "both statistically most common and socially preferable; it is the average and also an ideal."[141] In the Kinsey studies, the "normal" subject is manifested in its statistical ubiquity, but continually appears with its set of outliers. In this sense, averages do not exclude, but rather are connected to every entity in the studied group. Kinsey, Masters and Johnson demonstrate in their research that while average and norm are certainly co-constituted in the genealogy of sexual appetite, they are inextricable from *technique*. Techniques of measurement and observation condition the knowledge that emerges from those studies.

Masters and Johnson went on to have a prolific career in sex and marital therapy. Their manual of sex therapy, *Human Sexual Inadequacy*, outlined

[137] Cryle and Stephens, *Normality*, 343.

[138] Robyn Wiegman and Elizabeth A. Wilson, "Introduction: Antinormativity's Queer Conventions," *Differences: A Journal of Feminist Cultural Studies* 26, no. 1 (2015): 15.

[139] Janet Halley, *Split Decisions: How and Why to Take a Break from Feminism* (Princeton: Princeton University Press, 2006), 121.

[140] Cryle and Stephens, *Normality*.

[141] Stephens, "Normal," 143.

the principles and procedures of couple therapy, including how to take the patient's history and how to approach what they called the "marital unit" in a sensitive and dispassionate manner. *Human Sexual Inadequacy* also documented the major sexual dysfunctions, and several would ten years later appear in the third edition of the *Diagnostic and Statistical Manual of Mental Disorders* (DSM-III). These included problems of desire, ejaculation, orgasm, pain and "sexual inadequacy in the aging."[142] *Human Sexual Response* and their later guide on sex therapy cemented Masters' and Johnson's methodological approach to sexuality and its problems. The work of Kinsey and his team of researchers helped pave the way for Masters and Johnson by presenting data based on a very large sample and bolstering sexuality in the headlines, making it an important social issue. By the late 1970s, sex and marital therapy was a flourishing business in the United States. The focus on the functions and dysfunctions of sexual organs continues to preoccupy contemporary scientists. In addition, research focused on statistics and prevalence, similar to the Kinsey studies, is now very common in Western countries.

This chapter has canvassed the uses of techniques of calculation, measurement and observation in research on sexuality in the twentieth century in the works of the Kinsey team and Masters and Johnson. The Kinsey team utilised statistics and the concept of averages in order to produce an image of sexuality. The tables of data and graphs published in the *Male* and *Female* volumes served to develop not only an understanding of "average" sexual behaviour but also the prevalence of particular practices and certain "kinds" of individuals, such as the homosexual. For Kinsey, the statistical presentation of information on sexual matters was a means to expose the truth.[143] For him, scientists have a right to investigate as much as individuals have the right to know: "The scientist who investigates sexual behavior seems under especial obligation to make his findings available to the maximum number of persons, for there are few aspects of human biology with which more persons are more often concerned."[144]

Kinsey believed that society would benefit from accessing objective information on sexual behaviour. The drive to acquire (and publish) the truths of sexuality and the goal of presenting objective data connect Kinsey to Masters and Johnson. The latter likewise devoted their work to access-

[142] Masters and Johnson, *Human Sexual Inadequacy*, x. See also 316–350.
[143] Kinsey et al., *Female*, 9.
[144] Ibid., 11.

ing and publishing the facts of sexuality purportedly without social and cultural obstructions. By removing bodies from their everyday and placing them in the laboratory, Masters and Johnson sought to let bodies "speak" and merely observe and record information. Their research produced norms of physiological responses within the depths of bodies. Norms of sexual appetite in this context were also ideal responses of bodies in a sexual situation. Much like Kinsey's averages then, the techniques employed by Masters and Johnson as well as their interpretation of data conditioned the development of norms pertaining to sexuality. Not only are the norms of sexuality connected to the well-documented biases of Masters and Johnson, they are also intimately involved with the very techniques used to catalogue them.[145]

This chapter has demonstrated how the works of Masters and Johnson, and that of the Kinsey team, intensified the emergence of the calculable subject whose sexual appetite can be represented through numbers, charts and graphs, and can be compared to averages and norms. In the works of Masters and Johnson in particular, the question of "how much?" functioned at the centre of the human sexual response cycle, a concept which has remained foundational to the medicalisation of sexual appetite in the twentieth and twenty-first centuries. In 1980, the human sexual response cycle became a centrepiece of the new "Psychosexual Disorders" section in the DSM-III. Although disorders such as frigidity and nymphomania had featured in the DSM since its first edition in 1952, the DSM-III marks an enhanced turn to the pathologisation of the reduction of sexual appetite. The next chapter examines the discursive expansion in the classifications and interpretations of sexual dysfunctions in the DSM in the late-twentieth and early-twenty-first centuries. It approaches the development of the diagnostic manual as a technical object for both professional diagnosis and for the care of the self.

Bibliography

Bement Davis, Katharine. 1929. *Factors in the Sex Life of Twenty-Two Hundred Women*. New York: Harper & Brothers.

Bergler, Edmund. 1948. The Myth of a New National Disease: Homosexuality and the Kinsey Report. *The Psychiatric Quarterly* 22 (1–4): 66–88.

[145] Leonore Tiefer, *Sex is Not a Natural Act and Other Essays* (New York: Westview Press, 2004).

———. 1956. *Homosexuality: Disease or Way of Life?* New York: Hill and Wang.
Bullough, Vern L. 1994. *Science in the Bedroom: A History of Sex Research.* New York: Basic Books.
———. 2004. Sex Will Never Be the Same: The Contributions of Alfred C. Kinsey. *Archives of Sexual Behavior* 33 (3): 277–286.
Canguilhem, Georges. 1991 [1943/1966]. *The Normal and the Pathological.* Trans. Carolyn R Fawcett. New York: Zone Books.
Carlston, Erin G. 1997. 'A Finer Differentiation': Female Homosexuality and the American Medical Community, 1926–1940. In *Science and Homosexualities,* ed. Vernon A. Rosario, 177–196. New York: Routledge.
Cartwright, Lisa. 1995. *Screening the Body: Tracing Medicine's Visual Culture.* Minnesota: University of Minnesota Press.
Chauncey, George, Jr. 1989. From Sexual Inversion to Homosexuality: The Changing Medical Conceptualization of Female 'Deviance'. In *Passion and Power: Sexuality in History,* eds. Kathy Peiss and Christina Simmons, 87–117. Philadelphia: Temple University Press.
Cryle, Peter. 2010. The Average and the Normal in Nineteenth-Century French Discourse. *Psychology & Sexuality* 1 (3): 214–225.
Cryle, Peter, and Lisa Downing. 2010. Introduction: The Natural and the Normal in the History of Sexuality. *Psychology & Sexuality* 1 (3): 191–199.
Cryle, Peter, and Elizabeth Stephens. 2017. *Normality: A Critical Genealogy.* Chicago: University of Chicago Press.
D'Emilio, John, and Estelle B. Freedman. 2012 [1988]. *Intimate Matters: A History of Sexuality in America.* 3rd ed. New York: Harper & Row.
Datson, Lorraine, and Peter Gallison. 2007. *Objectivity.* New York: Zone Books.
Dickinson, Robert Latou, and Lura Beam. 1934. *The Single Woman: A Medical Study in Sex Education.* Philadelphia: The Williams & Wilkins Company.
Drucker, Donna J. 2010. Male Sexuality and Alfred Kinsey's 0–6 Scale: Toward 'A Sound Understanding of the Realities of Sex'. *Journal of Homosexuality* 57 (9): 1105–1123.
———. 2012. 'A Most Interesting Chapter in the History of Science': Intellectual Responses to Alfred Kinsey's *Sexual Behavior in the Human Male. History of the Human Sciences* 25 (1): 75–98.
———. 2013. Keying Desire: Alfred Kinsey's Use of Punched-Card Machines for Sex Research. *Journal of the History of Sexuality* 22 (1): 105–125.
———. 2014. *The Machines of Sex: Research Technology and the Politics of Identity, 1945–1985.* Dordrecht: Springer.
Faderman, Lillian. 2012 [1991]. *Odd Girls and Twilight Lovers: A History of Lesbian Life in Twentieth-Century America.* New York: Columbia University Press.
Foucault, Michel. 1976. *Histoire de la sexualité 1: La volonté de savoir.* France: Gallimard.

———. 1978. *The Will to Knowledge: The History of Sexuality, Volume 1*. Trans. Robert Hurley. London: Penguin Books.

———. 1994 [1973]. *The Birth of the Clinic: An Archaeology of Medical Perception*. Trans. A.M. Sheridan Smith. New York: Vintage Books.

Gebhard, Paul H., and Alan B Johnson. 1979. *The Kinsey Data: Marginal Tabulations of the 1938–1963 Interviews Conducted by the Institute for Sex Research*. Indianapolis: Indiana University Press.

Gerhard, Jane. 2001. *Desiring Revolution: Second-Wave Feminism and the Rewriting of American Sexual Thought 1920 to 1982*. New York: Columbia University Press.

Hacking, Ian. 2007. Kinds of People: Moving Targets. *Proceedings of the British Academy* 151: 285–318.

Halley, Janet. 2006. *Split Decisions: How and Why to Take a Break from Feminism*. Princeton: Princeton University Press.

Hamilton, George V. 1948 [1929]. *Research in Marriage*. New York: Lear.

Henry, George William. 1941. *Sex Variants: A Study in Homosexual Patterns*. New York: Paul Hoeber & Sons.

Hite, Shere. 1976. *The Hite Report: A National Study of Female Sexuality*. New York: Seven Stories Press.

Igo, Sarah E. 2007. *The Averaged American: Surveys, Citizens, and the Making of a Mass Public*. Cambridge, MA: Harvard University Press.

Irvine, Janice. 2005. *Disorders of Desire: Sex and Gender in Modern American Sexology*. Philadelphia: Temple University Press.

Jagose, Annemarie. 2013. *Orgasmology*. Durham and London: Duke University Press.

Jones, James H. 1997. *Alfred C. Kinsey: A Life*. New York: W. W. Norton & Company.

Katz, Jonathan Ned. 1995. *The Invention of Heterosexuality*. New York: Dutton.

Kinsey, Alfred C., Wardell B. Pomeroy, and Clyde E. Martin. 1948. *Sexual Behavior in the Human Male*. Philadelphia and London: W.B. Saunders Company.

Kinsey, Alfred C., Wardell B. Pomeroy, Clyde E. Martin, and Paul H. Gebhard. 1949. *Concepts of Normality and Abnormality in Sexual Behavior*. New York: Grune & Stratton.

———. 1953. *Sexual Behavior in the Human Female*. Philadelphia and London: W.B. Saunders Company.

Larousse, Pierre. 1874. *Grand dictionnaire universel du XIXe siècle (1866–1877), Tome onzième*. Paris: Administration du Grand Dictionnaire Universel.

Leeming, Anna, and Paul Brown. 1992. An Eclectic or Integrative Approach to Sex Therapy? *Sexual and Marital Therapy* 7 (3): 283–293.

Lloyd, Elisabeth Anne. 2005. *The Case of the Female Orgasm: Bias in the Science of Evolution*. Cambridge, MA: Harvard University Press.

Maier, Thomas. 2009. *Masters of Sex: The Life and Times of William Masters and Virginia Johnson, the Couple Who Taught America How to Love*. New York: Basic Books.

Masters, William H., and Virginia E. Johnson. 1966. *Human Sexual Response*. Boston: Little, Brown and Company.

———. 1970. *Human Sexual Inadequacy*. Boston: Little, Brown and Company.

———. 1974. *The Pleasure Bond: A New Look at Sexuality and Commitment*. Boston: Little, Brown and Company.

———. 1979. *Homoseuality in Perspective*. Boston: Little, Brown and Company.

———. 1994. *Heterosexuality*. New York: Harper Collins.

Playboy Magazine. 1968. Playboy Interview: Masters and Johnson. *Playboy Magazine*, May, pp. 67–82, 194–202.

Porter, Theodore M. 1994. Making Things Quantitative. *Science in Context* 7 (3): 389–407.

———. 1995. *Trust in Numbers: The Pursuit of Objectivity in Science and Public Life*. New Jersey: Princeton University Press.

Power, Michael. 2004. Counting, Control and Calculation: Reflections on Measuring and Management. *Human Relations* 57 (6): 765–783.

Reumann, Miriam G. 2005. *American Sexual Character: Sex, Gender, and National Identity in the Kinsey Reports*. Berkeley and Los Angeles: University of California Press.

Rose, Nikolas. 1988. Calculable Minds and Manageable Individuals. *History of the Human Sciences* 1 (2): 179–200.

———. 1998. *Inventing Our Selves: Psychology, Power, and Personhood*. Cambridge: Cambridge University Press.

Roubaud, Félix. 1855. *Traité de l'impuissance et de la stérilité chez l'homme et la femme comprenant des moyens recommandés pour y remédier*. Paris: J B Ballière.

Shulman, Alix. 1971. Organs and Orgasms. In *Women in Sexist Society: Studies in Power and Powerlessness*, eds. Vivian Gornick and Barbara K. Moran, 292–303. New York: Signet Books.

Stephens, Elizabeth. 2014. Normal. *TSQ: Transgender Studies Quarterly* 1 (1–2): 141–145.

Terry, Jennifer. 1999. *An American Obsession: Science, Medicine, and Homosexuality in Modern Society*. Chicago and London: University of Chicago Press.

The Eugene Register-Guard. 1950. Normal Sexual Behavior? No Such Thing Scientists. *The Eugene Register-Guard*. January 2, p. 2A.

Tiefer, Leonore. 1991. Historical, Scientific, Clinical and Feminist Criticisms of 'The Human Sexual Response Cycle' Model. *Annual Review of Sex Research* 2 (1): 1–23.

———. 2004. *Sex Is Not a Natural Act and Other Essays*. New York: Westview Press.

Trabsky, Marc. 2019. *Law and the Dead: Technology, Relations and Institutions*. Abingdon: Routledge.

Van de Velde, Theodore Hendrik. 1940 [1926]. *Ideal Marriage: Its Physiology and Technique*. London: William Heinemann.

Warman, Caroline. 2010. From Pre-normal to Abnormal: The Emergence of a Concept in Late Eighteenth-Century France. *Psychology & Sexuality* 1 (3): 200–213.

Waters, Chris. 2006. Sexology. In *Palgrave Advances in the Modern History of Sexuality*, eds. H.G. Cocks and Matt Houlbrook, 41–63. New York: Palgrave Macmillan.

Wiegman, Robyn, and Elizabeth A. Wilson. 2015. Introduction: Antinormativity's Queer Conventions. *Differences: A Journal of Feminist Cultural Studies* 26 (1): 1–25.

CHAPTER 5

The Diagnostic Manual and Technologies of Psychiatry

In May 2013, the American Psychiatric Association (APA) published the fifth edition of the *Diagnostic and Statistical Manual of Mental Disorders* (DSM-5). The DSM-5 introduces important modifications to the listings of sexual dysfunctions. First, while the human sexual response cycle (HSRC) is now removed from the manual, the DSM-5 nonetheless retains all the disorders that derive from this system of thought, with some alterations. Second, the DSM-5 alters classifications of low sexual desire and lack of sexual arousal in women. The sexual dysfunctions were originally elaborated around concepts that have long preoccupied sexology: arousal, desire, penetration, pain, pleasure and orgasm. Until the publication of the DSM-5, all the previous manuals contained variations of the phrase: "A Sexual Dysfunction is characterized by a disturbance in the processes that characterize the sexual response cycle."[1] From the DSM-III (1980) to the DSM-IV-TR (2000), issues of desire and arousal in women were dealt with via two separate disorders. In 2013, however, one disorder remains: Female Sexual Interest/Arousal Disorder (FSI/AD). In the following pages, this genealogy of sexual appetite will demonstrate that the deletion of "desire," and its substitution with "arousal" and "interest," represents an intensification of the problematisation of sexual imbalance in the diagnostic manual. In addition, I argue that the historical development of the DSM reveals an important refiguring in medical knowledge

[1] American Psychiatric Association, *Diagnostic and Statistical Manual of Mental Disorders*, 4th ed. (Washington, DC: American Psychiatric Association, 1994), 493.

J. Flore, *A Genealogy of Appetite in the Sexual Sciences*,
https://doi.org/10.1007/978-3-030-39423-3_5

from the pathologisation of sexual object choice to the problematisation of amounts, degrees and appetites in sexual relations.

The DSM-5 introduces several changes to the categorisation of psychiatric disorders, and similar to its predecessors, its release was met with controversy.[2] Despite criticism levelled against previous editions, several categories of sexual dysfunction originally introduced in the DSM-III remain in the DSM-5, albeit with changes to nomenclature and interpretation. In fact, the sexual disorders found in the DSM-5 owe much to the intellectual climate that pervaded psychiatry and sexology in the late twentieth century. The DSM-III foregrounded the question of sexual amounts and classified low sexual desire and Inhibited Sexual Excitement as paradigms of "psychosexual dysfunctions." The DSM-5 intensifies this turn to the problematisation of sexual appetite and is accompanied by modified techniques of diagnosis.

Reductions in levels of sexual interest, a lack of interest and an excessive appetite have increasingly become conspicuous foci of the diagnostic manuals. This chapter historicises the discursive expansion in the classifications and interpretations of "sexual dysfunctions" in contemporary US psychiatry in the late twentieth and early twenty-first centuries. It examines how the turn of the psychiatric gaze towards amounts, degrees and appetites in the DSM harnesses a range of techniques that, in addition to pathologising the patient, invite *her* to develop greater awareness of sexual subjectivity. The patient is actively encouraged to use the manual as well as information gathered during the therapeutic process as techniques for managing sexual imbalances. This chapter thus aims to historicise the turn of the psychiatric gaze towards the problematisation of sexual appetite in the late twentieth and early twenty-first centuries, and, in doing so, to approach the development of the diagnostic manual as a technical object for both professional diagnosis and for the care of the self.[3]

[2] See, for example, Arline Kaplan "DSM-V Controversies," *Psychiatric Times* 26, no. 1 (2009): 5–10 and Allen Frances, *Saving Normal: An Insider's Revolt against Out-of-Control Psychiatric Diagnosis, DSM-5, Big Pharma, and the Medicalization of Ordinary Life* (New York: William Morrow, 2013).

[3] The DSM is primarily used in the United States, though several countries also utilise it. Beyond the DSM, the *International Classification of Diseases*, now in its tenth edition, also contains a section dedicated to sexual disorders. The ICD provides standardised codes for diseases and enables the compilation of epidemiological data. It outlines few procedures for diagnosis in its section on sexual dysfunctions, limiting itself to definitions. It lists "loss of sexual desire" with frigidity and Hypoactive Sexual Desire Disorder both included and

The DSM, it should be emphasised, is not necessarily reflective of a consensus on the meanings and conceptualisations of sexuality and sexual appetite. However, my analysis of the DSM is informed by its status as an *index*. As Rachel Cooper notes, the DSM can be considered a "contact language," meaning that the manual provides a common vocabulary for health professionals.[4] In a constellation of varied symptoms, the DSM enables communication on diagnosis, prognosis and treatment. Given its political and cultural history, the DSM can be viewed as a repository of knowledge on sexuality; it reveals broader cultural concerns around sexuality, amounts and balance—hence its importance for the genealogy of sexual appetite.

The Role of Desire in the DSM-III

The DSM-III (1980) represented a break with the psychoanalytic tradition that had characterised US psychiatry since the DSM-I (1952). Earlier versions of the DSMs, as well as their predecessor, the military manual *Medical 203* (1946), conceptualised psychiatric symptoms as signs of a disturbed subconscious.[5] The DSM-III was published with the intention to efface psychoanalysis from the manual, firmly ensconce biological knowledge of the human mind, and produce classifications based on "shared clinical features."[6] Repudiating the subconscious, the DSM-III

"excessive sexual drive" with nymphomania and satyriasis named. The DSM has long abandoned the labels provided in the ICD. This applies to sexual orientation too, as the ICD still considers sexual orientation a determinant in certain disorders. For example, Sexual Relationship Disorder: "The gender identity or sexual orientation (heterosexual, homosexual, or bisexual) is responsible for difficulties in forming or maintaining a relationship with a sexual partner." Physicians and activists have called for sexual orientation to be removed from future versions of the manual; see http://www.who.int/bulletin/volumes/92/9/14-135541/en/. In the ICD, the question of appetite remains pervasive as disorders such as fetishism, lack or loss of sexual desire, excessive sexual drive and sexual aversion are organised around amounts and arousal.

[4] Rachel Cooper, *Psychiatry and the Philosophy of Science* (Montreal and Kingston: McGill-Queen's University Press, 2007), 94–95. See also Nikolas Rose, *Our Psychiatric Future: The Politics of Mental Health* (Cambridge: Polity, 2019), 71.

[5] The *Medical 203* was published in 1946 by the War Department and was specifically designed for use in the army. See Arthur C. Houts, "Fifty Years of Psychiatric Nomenclature: Reflections on the 1943 War Department Technical Bulletin, Medical 203," *Journal of Clinical Psychology* 56, no. 7 (2000): 935–967.

[6] Robert L. Spitzer and Janet B. W. Williams, "The Revision of the DSM-III," *Psychiatric Annals* 13, no. 10 (1983): 808.

emphasised categories of illness and deployed a revised system of classification.[7] For supporters of the paradigm shift from an "old psychiatry [derived] from theory [to] the new psychiatry [derived] from fact,"[8] the DSM-III embodied "science in the service of healing."[9] The third volume of the psychiatric handbook inaugurated the modern DSMs by transforming the manual from a "psychoanalytic-personality-development model to a more amorphous descriptive model with biological undertones."[10] Melvin Sabshin, medical director of the APA from 1974 to 1997, noted that the DSM-III and DSM-III-R (1987), while broad in scope, "attempt[ed] to provide objective criteria for diagnosing each disorder" and represented the "predominance of science over ideology."[11]

One of the hallmarks of the DSM-III was the establishment of the "multiaxial classification system." This technique was implemented to ensure that patients would be assessed in different domains and that physicians would not overlook key information. The first three axes constituted the "official diagnostic assessment,"[12] while Axes 4 and 5 took addressed psychosocial aspects of the individual's life and their capacity to engage with, or adapt to, social activities.[13] This approach to mental illness advocated a consideration of multiple aspects of the person's life, alongside the specific concern of the patient. Applying this methodology to sexuality, the DSM-III mobilised sexological principles and, in addition to disorders of sexuality based on biological functionality and inhibitions, incorporated Masters and Johnson's "complete sexual response cycle," a four-part system that delineated the physiological processes involved in a predominantly

[7] Rick Mayes and Allan V. Horwitz, "DSM-III and the Revolution in the Classification of Mental Illness," *Journal of the History of Behavioral Sciences* 41, no. 3 (2005): 249–275.

[8] Jerrold S. Maxmen, *The New Psychiatry: How Modern Psychiatrists Think about Their Patients, Theories, Diagnoses, Drugs, Psychotherapies, Power, Training, Families, and Private Lives* (New York: William Morrow & Company, 1985), 31.

[9] Gerald L. Klerman, George E. Vaillant, Robert L. Spitzer and Robert Michels, "A Debate on DSM-III: The Advantages of DSM-III," *American Journal of Psychiatry* 141, no. 4 (1984): 541.

[10] Houts, "Fifty Years of Psychiatric Nomenclature," 947.

[11] Melvin Sabshin, "Turning Points in Twentieth Century Psychiatry," *American Journal of Psychiatry* 147, no. 10 (1990): 1272. Note that the DSM-III was also designed to closely align with the World Health Organization's *International Classification of Diseases*, volume 9. See American Psychiatric Association, *Diagnostic and Statistical Manual of Mental Disorders*, 3rd ed. (Washington, DC: American Psychiatric Association, 1980), 399–457.

[12] APA, DSM-III, 23.

[13] Ibid., 23–32.

heterosexual encounter. In the DSM-III, the four stages of the cycle consisted of "appetitive, excitement, orgasm and resolution."[14]

The psychosexual dysfunctions found in the DSM-III could occur at any stage of the sexual response cycle. The manual classified reduction in levels of sexual excitement as a problem. Those fluctuations of intensity, which were depicted as signs of pathological sexual imbalance, would go on to characterise the modern DSMs' approach to the sexual dysfunctions. Indeed, the psychiatric gaze turned its attention in the DSM to questions of quantity and whether the patient was experiencing too little desire or too little excitement. That being said, the sexual appetites had already made their appearance in psychiatric nomenclature since 1952. In the DSM-I, frigidity and impotence were listed in Appendix C as supplementary terms of the urogenital system, while nymphomania featured as a condition affecting the psyche and the body, and yet did not impact "a particular system exclusively."[15] The DSM-II, on the other hand, only listed impotence as a "psychophysiologic" disturbance in which "emotional factors play a causative role."[16] Hence, while the DSM-III certainly introduced more disorders and revised existing typologies, the intellectual and medical terrain covered by the third edition of the manual was generally similar to that of its predecessors.[17] The presence of frigidity, impotence and nymphomania in appendices of the DSM-I and DSM-II highlights how the turn of the psychiatric gaze towards problems of quantity and intensity took place gradually over the second half of the twentieth century and subsequent revisions of the manual.

While the concern with the quantity and intensity of sexual activity was included in the DSM-I and the DSM-II, the incorporation of a description of mechanisms of sexual activity was a new addition. Masters and Johnson, as discussed in the previous chapter, pioneered the principles of twentieth-century American sex therapy. They expounded ideas of natural human function and asserted that they "[put] sex back into its natural

[14] Ibid., 276.

[15] American Psychiatric Association, *Diagnostic and Statistical Manual of Mental Disorders*, 1st ed. (Washington, DC: American Psychiatric Association, 1952), 120.

[16] American Psychiatric Association, *Diagnostic and Statistical Manual of Mental Disorders*, 2nd ed. (Washington, DC: American Psychiatric Association, 1968), 47.

[17] Mayes and Horwitz, "DSM-III and the Revolution in the Classification of Mental Illness," 251.

context."[18] Focusing strongly on patterns of behaviour and inhibitions, Masters and Johnson affirmed that all individuals possess the capacity for a satisfying sex life and they can help excavate this natural spring of satisfying desire, performance and pleasure. They posited that the four phases of the human sexual response were the same in both sexes, thus overthrowing "the myth of male sexual superiority"[19] and emphasising the female body's potential for sexual pleasure.[20] However, as critics have noted, the cure for Masters and Johnson was orgasm, not transformation of sociocultural mores. The sexologists failed to address heterosexism and gender diversity, and discarded gender differences as operating factors in sexual activity.[21]

Masters and Johnson particularly influenced Helen Singer Kaplan, an American physician and sex therapist whose work in the late 1970s was key to the design of sexual dysfunctions in the DSM-III.[22] Kaplan synthesised and modified Masters and Johnson's four-part sexual response cycle into a triphasic model of human sexuality: desire, excitement and orgasm.[23] Her work should therefore be placed within the context of late-twentieth-century medical discourses on female sexuality, which emphasised pleasure and orgasm.[24] In *Disorders of Desire and Other Concepts and Techniques in Sex Therapy* (1979), Kaplan introduced the concept of Inhibited Sexual Desire (ISD) into sexological discourse, which would only a year later become part of the DSM-III. Kaplan, Lynne Segal notes, used the "desire" phase of the response cycle to "bring in individual psychopathology to

[18] William H. Masters and Virginia E. Johnson, *Human Sexual Inadequacy* (Boston: Little, Brown & Company, 1970), 9.

[19] William H. Masters and Virginia E. Johnson, *Heterosexuality* (New York: Harper Collins, 1994), 42.

[20] Their work also documented women's ability to achieve multiple orgasms and this is repeated on page 276 of the DSM-III.

[21] See Janice Irvine, *Disorders of Desire: Sexuality and Gender in Modern American Sexology* (Philadelphia: Temple University Press, 2005). See also chapter three, "The Coital Imperative" in Lynne Segal, *Straight Sex: Rethinking the Politics of Pleasure* (Berkeley: University of California Press, 1994), 70–116.

[22] Leslie R. Schover, Jerry M. Friedman, Stephen J. Weiler, Julia R. Heiman and Joseph LoPiccolo, "Multiaxial Problem-Oriented System for Sexual Dysfunctions," *Archives of General Psychiatry* 39 (1982): 615.

[23] Helen Singer Kaplan, *Disorders of Desire and Other Concepts and Techniques in Sex Therapy* (New York: Simon & Schuster, 1979), 9–23.

[24] See John D'Emilio and Estelle B. Freedman, *Intimate Matters: A History of Sexuality in America*, 3rd ed (New York: Harper & Row. 2012 [1988]), 301–343.

help explain the rapidly growing 'disorders of desire.'"[25] Thus, despite her commitment to "pure" biology, Kaplan restored the importance of the psyche through the introduction of the word "desire" into the pathological condition and emphasised that the psyche was inhibiting the body's sexual potential. For Kaplan, ISD referred to "those situations of abnormally low libido in which an etiologic diagnosis has been made, that is, when it is established that sexual desire is inhibited by psychic factors."[26] As Segal remarks, "A disorder of desire might ... be attributed by Kaplan not just to 'ignorance', but to 'unconscious hostility', but would still be dealt with if, and only if, it distracted from successful sexual functioning."[27]

Kaplan considered that sexual functioning was the key to satisfying relationships and hence elided gender and sexual diversity, tensions in the domestic sphere and cultural differences as factors affecting sexual performance and sexual satisfaction. While low sexual desire had been an enduring concern in the history of sexual medicine, the first appearances of ISD are found in the works of two sexologists in 1977: Kaplan and sex therapist Harold Lief.[28] Both were members of the DSM-III Task Force for Psychosexual Disorders, along with central figures of twentieth-century sexology such as John Money, Robert Spitzer and Robert Stoller.

Difficulties during the "appetitive" phase of the sexual response cycle were covered in Inhibited Sexual Desire (ISD), a disorder defined as "Persistent and pervasive inhibition of sexual desire."[29] The diagnostic criteria of ISD stated that the clinician must take into account the "intensity and frequency of sexual desire." In addition, the DSM-III specified, "this diagnosis will rarely be made unless the lack of desire is a source of distress to either the individual or his or her partner."[30] The introduction of ISD formalised the psychiatric concern with amounts and intensity in sexual life, and turned to the question of distress. With this conceptual gesture, the patient is invited to participate in the production of the narrative of disease and the formulation of diagnosis. The inclusion of desire also silenced gender differences. The definition of desire appeared as "fantasies about sexual activity and a desire to have a sexual activity."[31] Desire was

[25] Segal, *Straight Sex*, 101.
[26] Kaplan, *Disorders of Desire*, 58.
[27] Segal, *Straight Sex*, 101.
[28] Harold Lief, "Inhibited Sexual Desire," *Medical Aspects of Human Sexuality* 7 (1977): 94–95.
[29] APA, DSM-III, 278.
[30] Ibid.
[31] Ibid., 276.

approached as an attraction or a kind of appeal in the psyche. One begins to fantasise and experiences an eroticised need. In the DSM-III, this need bears no gender distinction.

The remarkable break in the DSM-III was, first, the inclusion of the diagnostic category of Inhibited Sexual Excitement (ISE), and its separation from ISD. The DSM-III introduced ISE with the following note: "This has also been termed frigidity or impotence."[32] ISE was defined as "Recurrent and persistent inhibition of sexual excitement during sexual activity," with the following specification:

> *In males*, partial or complete failure to attain or maintain erection until completion of the sexual act, or
>
> *In females*, partial or complete failure to attain or maintain lubrication-swelling response of sexual excitement until completion of the sexual act.[33]

The difference in the experience of ISE between men and women was delineated in biological terms. It could be identified by examining lasting physiological changes, or their absence—vasocongestion in women, or muscular tension and vasocongestion in men. The reclassification of frigidity and impotence as ISE in the DSM-III reveals the medical interest in examining amounts in sexual activity, but specifically locating the ebb and flow of desire in biological processes, while the relevance of the psyche was located in ISD. Psyche and soma became distinct entities in the classification of ISE at least. However, the phase "Excitement" in the DSM-III included "a subjective sense of sexual pleasure" as a consideration alongside "accompanying physiological changes."[34] Pleasure nonetheless remained absent from the diagnostic criteria of ISE. The diagnostic criteria specify that the physician needs to judge whether "the individual engages in sexual activity that is adequate in focus, *intensity* and duration."[35] The physician has to inquire into how hard or how wet patients get, and in turn, the patients require an awareness of their body to provide an account of their body's receptivity and reactions.

The changes to the nomenclature of sexual dysfunctions in the DSM-III, which were made possible by the introduction of ISE and ISD, have

[32] Ibid., 279 (emphasis original).
[33] Ibid.
[34] Ibid., 276.
[35] Ibid., 279 (emphasis added).

not been as widely examined by historians of sexuality as the removal of homosexuality in 1973.[36] For example, Edward Shorter includes in his *Historical Dictionary of Psychiatry* an extensive discussion of homosexuality that includes its trajectory in the DSM. However, after a short section entitled "Homosexuality vanishes entirely from the DSM-III-R," Shorter turns to "Sex and gender identity disorder" without a discussion of the connections between object choice and sexual appetite in the compendium.[37] In a similar vein, Ronald Bayer's thorough account of the political climate of the DSM-III discusses arousal exclusively in relation to homosexuality without explicitly noting how this might reveal an enduring medical interest in the connections between appetite and object choice, or what this reveals about the epistemology of sexuality.[38] Yet, this interest in the associations and entanglements of object choice and sexual equilibrium was maintained by the DSM-III. For instance, while homosexuality was removed from the DSM in 1973, in 1980 the DSM-III retained the category "ego-dystonic homosexuality," which was dedicated to individuals whose homosexuality was a persistent source of distress. The description specified that "Individuals with this disorder may have either *no or very weak* heterosexual *arousal*."[39] The manual also remarked that "Individuals with Inhibited Sexual Desire may sometimes attribute the lack of sexual arousal to 'latent homosexuality.'"[40] This highlights that the degree or amount of arousal was still attached to object choice, as homosexuals could experience an imbalance of sexual arousal, that is, *too weak*, in a heterosexual context. With subsequent editions of the DSM, low sexual desire became further distanced from object choice.[41]

[36] Two important exceptions are Janice Irvine's *Disorders of Desire* and Leonore Tiefer, *Sex Is Not a Natural Act and Other Essays* (New York: Westview Press, 2004). The debates over homosexuality and its removal are described in Ronald Bayer, *Homosexuality and American Psychiatry: The Politics of Diagnosis* (Princeton: Princeton University Press, 1987 [1981]).

[37] Edward Shorter, *Historical Dictionary of Psychiatry* (Oxford: Oxford University Press, 2005), 131–132.

[38] Bayer, *Homosexuality and American Psychiatry*, 176.

[39] APA, DSM-III, 281 (emphasis added). Note that the ICD-10 lists "Egodystonic sexual orientation" but it bears no connection to levels of sexual appetite: "The gender identity or sexual preference (heterosexual, homosexual, bisexual, or prepubertal) is not in doubt, but the individual wishes it were different because of associated psychological and behavioural disorders, and may seek treatment in order to change it." See http://apps.who.int/classifications/icd10/browse/2016/en#/F66.1.

[40] APA, DSM-III, 282.

[41] Object choice of course continues to preoccupy science as demonstrated by the search for the "gay gene". See Simon LeVay, *Gay, Straight and the Reason Why: The Science of Sexual*

Works mapping the removal of homosexuality have seldom noted the connection between this change and sexual appetite. In an article on the DSM-III and the classification of mental illness, Mayes and Horwitz identify the removal of homosexuality as a key political event in the history of the manual.[42] However, there is no mention of how notions of "inhibited desire" or "inhibited excitement" flourished in the DSM and how connections were drawn between homosexuality and arousal. Likewise, Decker's *The Making of DSM-III*, Kirk and Kutchins' *The Selling of the DSM* and Bronski's *A Queer History of the United States* all cover the history of homosexuality in the manual, but devote little attention to the turn to sexual appetite.[43] In this historiography, sexual object choice remains a primary feature.

The DSM-III augmented the attention placed in the manual on questions of intensity and frequency. I argue that this marks the progressive emergence of discourses of sexual appetite and the idea of *balance* in sexuality. Thus, the final removal of ego-dystonic homosexuality from the DSM-III-R (1987) severed connections between object choice and amounts in the manual and expanded the problematisation of sexual appetite.[44] Contemporary psychiatric discourses, as exemplified by the DSM-III and its successors, now manifest a heightened concern with the problematisation of quantity; an individual who desires too little suffers from a pathological imbalance. The transformations in the architecture of psychiatric taxonomy from the DSM-I to DSM-III formalised and institutionalised a broader cultural concern with sexual amounts and balance.

The transformations in more recent editions of the manual also shifted the responsibility of the physicians. The emphasis of the psychoanalytic influence on the design of the first DSMs bestowed the physician with the

Orientation (Oxford: Oxford University Press, 2010) and Robert Alan Brookey, *Reinventing the Male Homosexual: The Rhetoric and Power of the Gay Gene* (Bloomington, IN: Bloomington University Press, 2002).

42 Mayes and Horwitz, "DSM-III and the Revolution in the Classification of Mental Illness," 258–259.

43 See Hannah S. Decker, *The Making of the DSM-III: A Diagnostic Manual's Conquest of American Psychiatry* (Oxford: Oxford University Press, 2013), Stuart A. Kirk and Herb Kutchins, *The Selling of the DSM: The Rhetoric of Science in Psychiatry* (New Brunswick: Transaction, 1992) and Michael Bronski, *A Queer History of the United States: ReVisioning American History* (Boston: Beacon Press, 2011).

44 See American Psychiatric Association, *Diagnostic and Statistical Manual of Mental Disorders*, 3rd ed, revised (Washington, DC: American Psychiatric Association, 1987).

responsibility to identify, excavate and care for the disturbed unconscious. The manual devoted little space to how diagnosis should proceed. Instead, the onus was on the physician to analyse, delve into the mind and uncover the sources of distress. This changed with the advent of the DSM-III where the manual became more focused on diagnostic accuracy, and the responsibility for recognising and managing sexual dysfunction became shared between the clinician and the patient.

The Medicalisation of Sexual Appetites in the Diagnostic Manual

From 1980 to 2013, the diagnostic manual further modified the structure and content of the classification of sexual dysfunctions. Low sexual desire, previously Inhibited Sexual Desire, was transformed into Hypoactive Sexual Desire Disorder (HSDD) in the DSM-III-R (1987). It remained as such, with no distinction between men and women, until 2013 when it was changed to Female Sexual Interest/Arousal Disorder (FSI/AD) and Male Hypoactive Sexual Desire Disorder (Male HSDD). Inhibited Sexual Excitement became Female Sexual Arousal Disorder and Male Erectile Disorder in 1987, before the former was incorporated into FSI/AD and the latter was defined as Erectile Disorder in 2013. The creation of FSI/AD and Male HSDD in the DSM-5 introduces striking gender differences in the manual to the extent that women can experience "sexual interest" or "arousal," while men can exhibit "desire." The DSM-5 thus marks a significant shift in the language of sexual dysfunctions, particularly insofar as it constructs gendered accounts of the quality and quantity of sexual activity.

This change in nomenclature, however, reveals how each subsequent volume of the DSM has further encouraged patients to conduct themselves as responsible sexual subjects. In the DSM-5, the dissemination of the language of sexual balance, in conjunction with the application of measuring devices such as questionnaires and inventories, has intensified the role of the patient in sharing responsibility for diagnosing and managing their pathology. That is to say, the development of the DSM as a diagnostic manual for psychologists and psychiatrists in the twenty-first century has also produced techniques for patients to assume responsibility for managing a balanced diet of sex. Critiques of the changes in psychiatric approaches to sexuality have focused on orientations and "problematic" objects of desire such as the ones found in the paraphilias. In a special issue of *Archives of Sexual Behavior* on the DSM-5 and "Classifying Sex," for example, the authors focus almost exclusively on changes pertaining to object choice, the paraphilias and gender identity.[45] In contrast, I argue that an enduring, and increasingly more prominent, dimension of psychiatric knowledge remains unexamined, that is the concern with quantity and balance. A contributor to the special issue, Alain Giami, notes that the ICD and the DSM have abandoned moral references and instead turned to models of rights and responsibility in their classification of sexual disorders towards "a framework based on communication, individual freedom, well-being, and equality."[46] While this is undoubtedly the case, the focus on freedom and the responsibilisation of the subject also hinges upon a model of sexuality where *moderation* is central.

The expansion of the categories of psychiatric disorder, alongside a renewed emphasis on gender differences in appetite and desire, represents the minute medical detailing of the sexualised body. Object choice remains the subject of medical analysis, but not content with naming the desiring body as a site of problematic sexual behaviour, psychiatric knowledge on sexuality in the twenty-first century anatomises the body, considers its amounts and rhythms in more depth, and turns to female sexuality as a dyad between interest and arousal.

In Foucault's reading of Ancient Greek scholarship, which this book has discussed in more detail in Chap. 1, sexual relations were submitted to a regimen. Subjects assumed responsibility for managing their pleasures.[47] This mode of governance was concerned not so much with objects of

[45] See, for example, Lisa Downing, "Heteronormativity and Repronormativity in Sexological 'Perversion Theory' and the DSM-5's 'Paraphilic Disorder,'" *Archives of Sexual Behavior* 44, no. 5 (2015): 1139–1145, Alain Giami, "Between DSM and ICD: Paraphilias and the Transformation of Sexual Norms," *Archives of Sexual Behavior* 44, no. 5 (2015): 1127–1138, and Jeffrey Weeks, "Beyond the Categories," *Archives of Sexual Behavior* 44, no. 5 (2015): 1091–1097. See also by Jack Drescher, "Queer Diagnoses: Parallels and Contrasts in the History of Homosexuality, Gender Variance, and the *Diagnostic and Statistical Manual*," *Archives of Sexual Behavior* 39, no. 2 (2010): 427–460, and "The Removal of Homosexuality from the DSM: Its Impact on Today's Marriage Equality Debate," *Journal of Gay & Lesbian Mental Health* 16, no. 2 (2012): 124–135.

[46] Giami, "Between DSM and ICD," 1136.

[47] Michel Foucault, *The Use of Pleasure: The History of Sexuality, Volume 2*, trans. Robert Hurley (New York: Vintage, 1985), 40.

desire, rather the regimen of *aphrodisia* focused on the prudent management of the quality and quantity of bodily relations and rhythms. Excess and passivity "were the two main forms of immorality in the practice of the *aphrodisia*."[48] In fact, self-governance for the Greeks cannot be limited to what is understood in the twenty-first century by "sexuality" or even sexual relations, for the use of pleasures was conceived by the Greeks as "a whole art of living."[49] Foucault asserts that the use of pleasures of (free male) individuals at the time was marked by an attention to the stylisation of an ethics of the self. In other words, it "did not seek to justify interdictions," but to cultivate an aesthetics, an art of existence.[50] Ancient Greek writings on the development of an ethics or a regimen of eroticism focused on questions of dietetics, moderation and balance. They expressed concern for how one could use and regulate their pleasures and take care of their relations.[51] Far from being confined to individuals and bodies in isolation, recommendations on *sexual dietetics* were relational.

What we have then is the confluence of a tripartite discourse that forms the "techniques of the living"[52]: pleasures must be recognised, practised and managed; their management must aim towards achieving moderation and balance, neither too much nor too little; and finally, to properly manage pleasures, individuals must develop tools for cultivating an ethical life. The development of the practice of the care of the self constituted a lifelong work. Individuals were encouraged to perform an assessment of themselves, of their thoughts and practices to develop a regimen of moderation and a strategy for regulating their conduct in relation to others. The fundamental technique that Foucault identifies in ancient schools of thought, that of *Askesis*, generates a set of tools that involves the training of the mind, the body and the soul.[53] The practice of taking care of the self, then, involves a procedure of testing the self, which refers to a mode

[48] Ibid., 47. See chapter one of this book for a discussion of *aphrodisia*.

[49] Ibid., 101.

[50] Ibid., 97.

[51] Ibid., 55–56.

[52] Michel Foucault, "Subjectivity and Truth," in *Ethics: Essential Works of Foucault, 1954–1984, Volume 1*, ed. Paul Rabinow (London: Penguin Books, 1997), 89.

[53] Michel Foucault, "Technologies of the Self," in *Technologies of the Self: A Seminar with Michel Foucault*, eds. Luther H. Martin, Huck Gutman and Patrick H. Hutton (Amherst: University of Massachusetts Press, 1988), 239. See also Michel Foucault, *The Care of the Self: The History of Sexuality, Volume 3*, trans. Robert Hurley (New York: Vintage Books, 1986).

of self-evaluation, appraisal and reflection. It is a course of actions that the individual cultivates, performs and repeats in the formation of subjectivity.

The adaptation of Foucault's reading of Ancient Greco-Roman texts to contemporary contexts must be approached with caution. Certainly, the subjectivity of Ancient Greeks and Ancient Romans differs sharply from the modern Western era. However, as Eric Paras notes:

> To the extent that ethical questions were ones in which what was at stake was the way in which free individuals related to one another … the ancient arts of living—while not directly imitable—had the potential to speak to our situation.[54]

Reading Foucault's series on the history of sexuality might offer a critical vocabulary to examine the production of knowledge on subjectivity and sexual appetite in the twenty-first century, without claiming these concepts as directly applicable. As a result, rather than speaking of history in terms of grand ruptures and unprecedented inventions, the past might offer us *insight* into continuities and transformations in the present tense. Foucault argues that one develops ethical subjectivity through certain techniques of subjectification that involve testing, managing and improving the self. Although the content and techniques of ethical subjectivity have been transformed, what is continuous is how the concern with sexual appetites, and the techniques for their management and regulation, has endured in various forms. Through the development of a diagnostic manual and the psychiatric therapeutic encounter, the individual adopts different techniques and develops strategies for inspecting and managing the self. The objective of self-analysis, and what has never disappeared from sexual medicine, is an intense examination of the dietetics of pleasure and desire. Nikolas Rose extends Foucault's work by drawing links between the ancient practice of the care of the self and modern therapeutic processes. He observes that texts of psychotherapy operate as "a kind of instruction manual in the techniques of the self."[55] What also emerges from this system of knowledge, I suggest, is the concept of *balance*. In the twenty-first century, the subject needs to manifest and maintain equilibrium in various

[54] Eric Paras, *Foucault 2.0: Beyond Power and Knowledge* (New York: Other Press, 2006), 131.

[55] Nikolas Rose, *Governing the Soul: The Shaping of the Private Self* (London and New York: Routledge, 1989), 248.

facets of life, including sexuality. To achieve balance, the subject is armed with tools of diagnosis and techniques of self-management.

The DSM-5 supplies the clinician with guidelines and procedures that must be followed for ascertaining a diagnosis of sexual dysfunctions. Yet in the deployment of those criteria during the therapeutic encounter, patients are not only expected to reveal their sexual life, they are required to actively participate in the process of making diagnoses by examining the frequency, duration and intensity of their sexual practices. Since at least 1980, the DSM has emphasised that in the examination of a patient's sexuality, all aspects of their interpersonal relationships have to be considered. In the process of confessing to the clinician their pleasures, desires, relations and actions, patients need to recognise themselves as subjects of scientific inquiry. The late-twentieth and early twenty-first-century psychiatric approach to sexuality in the DSM requires the patient to become more active in the development of their diagnosis. Indeed, the manual emerges as a technique for managing sexual balance.

Disorders of sexual appetite emerge not only through clinical practices, but also through critical practices carried out by the patient. As several scholars have noted, it is problematic to assume that diseases are stable entities that can be discovered by the techniques of modern medicine.[56] By coming into contact with techniques of classification, the DSM partakes in enacting sexual dysfunctions and promotes an awareness of the different conditions. This is also reproduced in the work the patients perform on themselves, through the practice of self-monitoring and self-management. In other words, the patient and the psycho-medical techniques produce, debate and reconfigure the disorder. The self-aware subject is enacted and produced, repetitively, by using clinical techniques, such as those of classification and examination, which are harnessed to garner information about the sexual life of the patient.

Questionnaires of Disorder

In the DSM-5, FSI/AD is found in Section II: Diagnostic Criteria and Codes under "Sexual Dysfunctions." The technique of classification relies on different diagnostic criteria in the making of pathology and the manual specifies conditions that have to be met for a patient to be diagnosed with

[56] See, for example, Annemarie Mol, *The Body Multiple: Ontology in Medical Practice* (Durham and London: Duke University Press, 2002).

FSI/AD. The classification produces an idea, a picture of the patient of FSI/AD. The system of criteria delineates steps and procedures that must be followed by the physician in the performance of her/his role. At the same time, the criteria set forth objects, activities and circumstances on which the patient will need to comment. A classification system, Geoffrey C. Bowker and Susan Leigh Star explain, is "a set of boxes (metaphorical or literal) into which things can be put to then do some kind of work– bureaucratic or knowledge production."[57] In the assemblage of one's personal biography, the subject engages with the classification of the DSM and its instruments, questionnaires and technical equipment.

Criterion A of the FSI/AD requires patients to account for their patterns of sexual activity:

> A. Lack of, or significantly reduced, sexual interest/arousal, as manifested by at least three of the following:
>
> 1. Absent/reduced interest in sexual activity.
> 2. Absent/reduced sexual/erotic thoughts or fantasies.
> 3. No/reduced initiation of sexual activity, and typically unreceptive to a partner's attempts to initiate.
> 4. Absent/reduced sexual excitement/pleasure during sexual activity in almost all or all (approximately 75%–100%) sexual encounters (in identified situational contexts or, if generalized, in all contexts).
> 5. Absent/reduced sexual interest/arousal in response to any internal or external sexual/erotic cues (e.g., written, verbal, visual).
> 6. Absent/reduced genital or nongenital sensations during sexual activity in almost all or all (approximately 75%–100%) sexual encounter (in identified situational contexts or, if generalized, in all contexts).[58]

For all these indicators, the patient will need to provide certain specifications: whether the "condition" is lifelong or acquired,[59] whether she is experiencing disturbances that are generalised or situational; and with the help of the patient's narrative, the clinician will specify the level of distress:

[57] Geoffrey C. Bowker and Susan Leigh Star, *Sorting Things Out: Classification and Its Consequences* (Cambridge, Massachusetts: MIT Press, 1999), 10.

[58] American Psychiatric Association, *Diagnostic and Statistical Manual of Mental Disorders*, 5th edition (Washington, DC: American Psychiatric Association, 2013), 433.

[59] "Lifelong" means that the "disturbance has been present since the individual became sexually active" while "acquired" refers to a disturbance that "began a period of relatively normal sexual function." APA, DSM-5, 433.

mild, moderate or severe. Criterion A covers different concepts of sexual activity: interest, fantasies, initiation, pleasure, receptivity and genital functionality. The patient needs to be able to provide an account in response to all these different indicators. The different paradigms incorporate subjective sensations related to attraction as well as an evaluation of physiological responses. Criteria A1 and A2 ask the patient about her interest in sexual activity as well as her sexual or erotic thoughts and fantasies. To account for those two aspects, the patient must comment on whether there has been a reduction in her interest in sexual activity and whether she fantasises about having sex. She is required here to comment on the quantity and quality of interest and thoughts. The patient needs to provide an account of her emotional inclinations: does she imagine erotic and sexual scenarios; is she interested in sexual relations; does she exercise her mind erotically; and if there has been a reduction in carnal appeal, when did this begin?

Criterion A3 focuses on the question of initiation and receptivity. This introduces the examination of the patient's relations, or rather, how she is managing sexual activity with one, or more, partner/s. As the DSM-5 states, Criterion A3 is "behaviorally focused. ... A couple's beliefs and preferences for sexual initiation patterns are highly relevant to the assessment of this criterion."[60] In the application of this criterion, then, the patient will need not only to account for her own behaviour, but also to discuss how her partner reacts to her lack of initiation or her lack of responsivity. By taking into account her behavioural patterns and that of her partner, an "ideal" outcome is produced. That is to say, how she would like to act or respond when faced with a sexually charged situation. The patient thus actively fashions her own sexual subjectivity: she produces a picture of what her sexuality *could* become. For instance, she might desire to initiate sexual activity more, which could lead her to confess to her *distress* about her inability to initiate. Conversely, she may be distressed because she does not experience excitement when faced with sexual initiation. The next criterion, A4, delves into the experience of excitement and pleasure. Here the patient needs to give information about the situations and contexts where excitement and pleasure are decreased or absent. She needs to think about the past six months and consider whether the reductions or absences define most of her sexual relations during that time.

[60] Ibid., 434.

The diagnostic criteria require that patients scrutinise and monitor their bodily rhythms, relations and actions in order to recognise whether they fit at least three of the six criteria. I want to suggest here that, through the technique of the manual and by providing information on the different criteria, the patient is encouraged to participate in a project of enacting her subjectivity. Patients are tasked with engaging in self-entrepreneurialism where they educate themselves in developing the ability to recognise interest or erotic thoughts, and identify receptivity. In other words, the patient will develop an awareness of her mind, body and passions, and regulate any imbalances on different levels by self-monitoring and engaging in social intercourse armed with this knowledge. An analysis of FSI/AD reveals that the clinical examination demands that the patient be active, cooperative and contribute to the diagnosis. The clinical encounter is central to the deployment of sexuality; as the patient takes a seat in the therapeutic office, she is immersed in the "institutional incitement"[61] to speak about sex. There is no doubt that during this process, through the procedures delineated by the manual, the individual will provide an account of her feelings, interests and desires. However, in her interaction with the diagnostic criteria, the patient is far from a passive recipient of psychiatric expertise; in fact, the patient is already working on herself in the form of self-analysis.

Criterion A5 asks the patient to give an account of her responses to "sexual/erotic cues." She is asked here to comment on how erotic stimuli affect her interests and her arousals. The patient will need to be able to distinguish between her physical and subjective reactions, as well as provide commentary on her ability to make this distinction. She will also answer questions as to what form of stimuli was most or least effective on her senses. For example, she will reveal her consumption of erotic material, if any. Hence Criterion A demands that the patient comment on various paradigms. Criterion A6 however requires commentary on genital sensations, and similar to Criteria A4 and A5, the patient needs to provide information about contexts and circumstances. While the system of classification requires that patients comment on their subjective feelings and physiological responses, the manual is quite clear that it relies on the patient's *perception* of vaginal engorgement:

[61] Michel Foucault, *The Will to Knowledge: The History of Sexuality, Volume 1*, trans. Robert Hurley (London: Penguin, 1978), 18.

[Criterion A6] may include reduced vaginal lubrication/vasocongestion, but because physiological measures of genital sexual response do not differentiate women who report sexual arousal concerns from those who do not, the self-report of reduced or absent genital or nongenital sensations is sufficient.[62]

The patient must make an assessment of her genital responses, and how aware she is of lubrication and sensations in genitalia. The patient has the responsibility to *estimate* her viscous imbalance. She needs to be able to comment on whether her subjective awareness of desire, that is her interest, *matches* and is balanced with, her physiological changes.

Since 1987, the DSM has noted that symptoms must cause "clinically significant distress in the individual," in accordance with criterion C. The DSM-5 also states that distress "may be experienced as a result of the lack of sexual interest/arousal or as a result of significant interference in a woman's life and well-being."[63] The patient needs to account for her relations and comment on whether the "condition" impacts on her ability to maintain intimate relationships. She needs to comment on how her lack of sexual interest and arousal affects her personal and relational well-being. Like the other criteria, distress is a performative category. The patient will express her degree of anxiety and anguish, and simultaneously consolidate an account of sexual appetite as distressingly lacking. The psychiatrist will use those answers to specify the "current severity" of the patient's condition as either mild or moderate or severe. In giving an account of one's sexual history and one's distress, the patient is always already immersed in the vocabulary of self-evaluation. In this process, she turns to her sexual past and present in order to locate and diagnose a problem, and invests hope for change, for balance in her future self.

Under criterion B, FSI/AD specifies that symptoms in criterion A must have persisted "for a minimum duration of approximately 6 months."[64] Those requirements are important in order to rule out temporary or momentary sexual shortcomings and establish that there is a psychiatric condition. In the description of disease, and the naming and production of diagnosis, the DSM relies on behaviours, functions, signs, symptoms

[62] APA, DSM-5, 434.
[63] Ibid.
[64] Ibid., 433.

and data on the progression of disease over a period of time.[65] Thus, criterion B of FSI/AD stipulates that the symptoms must have persisted for about half a year. This indicator requires the individual to *keep time*. Patients need to monitor their imbalances in sexual appetite. They need to be attentive to the changes in amounts of desire, when and where they desired less, and whether the situation went on for a minimum of six months. Patients need, in a sense, to "write down" their desires and arousals and diarise them, to project their imbalances upon an organisation of duration *and* location.

The change in language from desire to interest draws attention to the problem of self-awareness in the therapeutic encounter. It highlights the difficulties for the subject in distinguishing between their desires and interests. In light of the removal of desire from female sexual dysfunctions, I would argue that women, in particular, need to be aware of the differences between their interests—that is, attraction in their experience of sexuality—and their arousals. This means that *any* sexual activity must be assessed and evaluated in terms of bodily responses. The patient will make an assessment of her genital sensations during the sexual relations that she engages in over the course of six months and she will compare the different occasions. The key feature here is that the reduction in the number of times one thinks about sex, responds to initiation or initiates sex becomes a problem only when it occurs frequently, and it causes distress. In addition, the DSM is interested in the circumstances of reduction in desire. Patients are incited to develop a certain awareness or alertness to their *rhythms of desire*. This is an imbalance which, should it cause clinically significant distress, requires clinical intervention. Thus, there is a dual action of management and regulation, while the onus remains on individuals to monitor their rhythms in order to comprehend their disorder. Finally, the last indicator of FSI/AD, Criterion D, ensures that the dysfunction is not the result of another mental disorder outside of the realm of sexuality, that the patient is not suffering from relationship distress and that her condition is not the result of another medical condition or the effect of medication.

During the passage through the clinic, the patient discovers, uncovers patterns of behaviour, emotional and physical truisms and fashions, with

[65] See further Ilina Singh and Nikolas Rose, "Biomarkers in Psychiatry," *Nature* 460 (2009): 202–207 and Rachel Cooper, "What's Wrong with the DSM?," *History of Psychiatry* 15, no. 1 (2004): 5–25.

the help of psychiatric language and tools, an "improved" self. The patient also develops techniques that assist in the continued management of the self. The classification technique is pedagogical and emancipatory. The patient will come out of the clinical process having learnt about certain events, patterns and behaviours. Caring for the self, then, is a process of developing and enacting subjectivity. It entails the awareness of what kind of subject one is and what kind of subject one *aspires* to be. This process does not tend towards finality, as the subject needs constant monitoring and improvement, and hence invokes a "becoming" of subjectivity.

The DSM's classification of FSI/AD deploys different criteria and requires the psychiatrist to be skilled in the interpretation of such paradigms in order to detect pathology. But the technique of classification in the DSM also provides a set of tools and instruments designed to make the modern subject accountable for her condition. When the patient answers questions about her interest in sexual relations, if her mind wants it, and if her body wants it, and whether the two are in harmony, and when she has to keep time, chart her patterns of interests and arousals, and keep a record of how she reacts to different erotic settings, she is being encouraged to take responsibility for her desires and her pathology. Patients must know what amounts of desire they are experiencing and whether that is too much or too little, how this affects their relationships and whether they experience distress. They also need to be able to chart the ebbs and flows of their arousals across a finite period of time. At the beginning of the twenty-first century, the psychiatric gaze is not only marked by a sustained attention to biology and its dysfunctions, it is also interested in the capacity of the body to lend itself to measurement and re-establish an idealised balance. Those functions need to be excavated and studied in order to attain the source of the disorder. This, however, is compounded with notions of individual responsibility, as Rose has extensively discussed.[66] Individuals not only need to provide an account of their desires and their intensity, but also need to take care of their bodies, moderate the use of pleasures and assume responsibility for their relations.

In the diagnosis of FSI/AD, the medical subject needs to confess desires, however perverse or immoral, to sketch a personalised, individual

[66] See by Rose, *Governing the Soul*, *Inventing Our Selves: Psychology, Power, and Personhood* (Cambridge: Cambridge University Press, 1998) and *The Politics of Life Itself: Biomedicine, Power, and Subjectivity in the Twenty-First Century* (Princeton: Princeton University Press, 2007).

sexual historiography. The confession requires a patient to have insight into her relations, rhythms and actions. To help the patient *think* about symptoms of her sexual imbalances, psychiatrists routinely make use of questionnaires as part of the process of diagnosis. The guidelines of the DSM are supported by questionnaires that delve into the intensity of the patient's sexual appetite and her levels of distress. In addition, the use of scales and inventories form an integral part of the clinical examination. They are designed to help the physician conduct the examination of the patient's condition:

> Finding the correct way to ask questions and to decode answers on sexual health and illnesses might be difficult and, in some way, embarrassing. Hence, expert-guided, validated and standardized sexual inventories … might help naive and more experienced physicians alike to address sexual health and diseases.[67]

Studies have shown that clinicians find it challenging to respond to complaints about a lack of sexual interest/arousal.[68] Hence questionnaires "encourage physicians to discuss sexuality as part of the routine patient encounter."[69] I suggest, however, that they not only support the work of the clinician, but also incite the patient to speak about sex, to reveal their desires and their intensities, and work to achieve what can be referred to as instances of self-recognition. The questionnaires encourage patients to evaluate, "take stock" of their own histories of sexuality. They prompt patients to *think* about sexual balance, amounts and frequencies in order to be able to respond to the questions. Through such exercises on the self, the patient accesses the techniques of self-diagnosis and participates in the formation of sexual subjectivity. The different techniques embedded in the manual, then, have a pedagogic function, as patients need to learn what behaviours are conducive to their sexual well-being.

[67] G. Corona, E.A. Jannini and M. Maggi, "Review: Inventories for Male and Female Sexual Dysfunctions," *International Journal of Impotence Research* 18, no. 3 (2006): 237.

[68] Veronica Harsh, Elizabeth Lloyd McGarvey and Anita H Clayton, "Physician Attitudes Regarding Hypoactive Sexual Desire Disorder in a Primary Care Clinic," *Journal of Sexual Medicine* 5 (2008): 640–645.

[69] Anita H Clayton et al., "Cutoff Score of the Sexual Interest and Desire Inventory-Female for Diagnosis of Hypoactive Sexual Desire Disorder," *Journal of Women's Health* 19, no. 12 (2010): 2191.

Self-Diagnosis and the Problem of Professional Expertise

The assessment of quality and quantity is not only conducted in relation to the intensity of sexual appetite. In contemporary psychiatry, the whole life of the modern subject needs to be studied. The DSM-5 advises physicians to pay attention to factors that relate to the patient's romantic relationships, their emotional state, for example, their body image and a history of emotional or sexual abuse, other psychiatric elements such as depression and anxiety, "stressors" such as bereavement and job loss, and their cultural and religious backgrounds. Reductions in sexual appetite are never dealt with in isolation. It is not just about self-regulation and the management of sexual appetite, but rather about the examination of a life. The subject needs to canvass their life, open it to a self-reflective evaluation and make connections between different events. How one conducts their life and manages those different features are relevant to the pronunciation of pathology. The diagnostic apparatus of the DSM operates only with the input of the patient. By cooperating, revealing and monitoring, the subject understands herself and thus operationalises techniques for an ethics of existence.

Through the psychiatric encounter, the patient is armed with tools of self-diagnosis. By "self-diagnosis," I do not mean that individuals absorb the discourse of the DSM. As I have emphasised earlier in the chapter, contemporary psychiatry requires and encourages the patient to become active in the making, unmaking and practice of disease. The use of the term "self-diagnosis" emphasises how the self-management of sexual appetite is crucial for the professional diagnosis of the disorder. Self-diagnosis results from the interrelation between the forensic analysis undertaken by the patient and their visit to the psychiatrist. The project of the self as enterprise requires that the patient "take on" the information gathered through the clinical venture. This is not solely achieved through diagnosis, but also attained by the patient pondering her own thoughts, actions and relationships. Self-diagnosis as an *ethics* of existence signifies the building or shaping of a life based, or drawing from, the psychiatric examination. As Rose observes, "[t]he therapeutic subject is destined to leave therapy and live their life; but the self-techniques of therapy are to accompany them always."[70] The tools will be applied continuously as the

[70] Rose, *Governing the Soul*, 247.

patient continues to self-monitor, but once this procedure is enacted, the patient is armed with techniques of self-diagnosis and instruments to fashion an ethical self. It is not so much that the individual adopts the pathology passively; instead, the individual comes to take care of herself *through* the knowledge of pathology.

The modification of categories of sexual dysfunctions in the DSM reveals a transformation in the DSM from a morality of unhealthy sexual practice to an ethics of accounting for oneself and to a call to take responsibility for one's own practices, at least as far as sexual dysfunctions are concerned. This emerges as ideas of balance, control and moderation feature more prominently as the categories pertaining to sexuality are refined. From 1987, the DSM's approach to sexual dysfunctions was entirely devoted to imbalances of sexual appetite. What we find is that the individual is not encouraged to cease certain practices. Rather she is encouraged to take responsibility and manage those desires. The instrument of classification, and the associated techniques of self-assessment and self-awareness, operationalises ideas of self-reliance and regimen. The pathologisation of sexual appetite through the technique of the diagnostic manual exceeds professional use, as the manual becomes a text for the fashioning of the self. What emerges from the use of the manual are interdependent discourses of self-diagnosis of sexual imbalance and the professional diagnosis of pathology. The complementary discourses arise from the application of the diagnostic manual and its instruments, and from the interaction between the patient and the psychiatrist.

The authority of the psychiatrist is also reinforced by the caveats located within the system of classification. The DSM produces categories of sexual disorders and aims to cover the different components of the sexual encounter. While the phases are divided between arousal, excitement and orgasm, the language of sexual appetite produced by the DSM includes residues or excesses of psychiatric language that it cannot fully incorporate. The vast possible fluctuations of sexual imbalances cannot be covered and contained in the instrument of the manual. Thus, from 1980 to 2000, the DSMs contained categories pertaining to psychosexual disorders "not otherwise specified" or "not elsewhere classified" to cater for the excesses of professional knowledge and to provide space for the physician to develop specific diagnoses. The DSM-5 includes "other specified sexual dysfunction" and "unspecified sexual dysfunction." These are disorders that do not meet all the requirements for the diagnosis of sexual dysfunction.

Those exceptions reflect the excesses of a language of pathology and serve to embed the expertise of the clinician into psychiatric nomenclature. When a sexual dysfunction is difficult to determine, or the physician chooses not to provide specification, professional expertise becomes primary and only she/he has the institutional legitimacy to engage in naming practices. The techniques of psychiatric classification alongside the use of the questionnaire encourage the patient to articulate and tailor her emotions, performance and moods in medical language and become more alert to the quality and quantity of her passions. Nonetheless the passage through the therapeutic process necessitates the continuing involvement of the clinician in monitoring or guiding the patient, and the DSM reinstates the subjectivity of the physician, bestowing upon them the final authority to name the "disorder."

This chapter has examined the development of the DSM and selected tools of diagnosis in the twentieth and twenty-first centuries. It has explored how ideas of amounts, moderation and balance circulate through the dissemination of the diagnostic manual and its techniques of classification and examination. The accentuation of discourses of sexual appetite occurred with the publication of the DSM-III in 1980. In this edition, the concern with object choice began to recede from view and psychiatry became increasingly preoccupied with notions of too little or too much appetite. This chapter has not delved in depth into the question of sexual excess. This is because excess has yet to find its standalone diagnosis in the DSM. However, as Irvine remarks, the problem of "sex addiction" emerged in North America around the same time ISD and ISE were introduced in the DSM.[71] In addition, the DSM-III contains under "psychosexual disorder not elsewhere classified … distress about a pattern of repeated sexual conquests with a succession of individuals who exist only as things to be used (Don Juanism and nymphomania)."[72] The DSM-III-R, DSM-IV and DSM-IV-TR all list similar explanations. This was

[71] See Irvine, *Disorders of Desire*, 163–183.

[72] APA, DSM-III, 283. However, disorders listed under "paraphilic disorders," for example, "sexual sadism disorder" and "pedophilic disorder," could be interpreted as "excessive." Excess as a "standalone" pathology through the proposed category "hypersexual disorder" has not been included in the DSM-5. I maintain that lack of sexual appetite is always in connection with both sexual lack and excess. For instance, the condition of low or absent sexual appetite can be considered excessive in itself since it represents the "pathological" exaggeration of a norm of sexuality, albeit towards lack.

removed from the DSM-5, although "hypersexual disorder" continues to preoccupy scientists and physicians outside the sphere of the manual.[73]

The emergence of discourses of sexual appetite occurred alongside the elaboration of pedagogical and emancipatory ideas of the care of the self. If the DSM maintains that sexual response "has a requisite biological underpinning, yet it is usually experienced in an intrapersonal, interpersonal and cultural context,"[74] then techniques such as classification, interviews, questionnaires and surveys work towards making the patient more aware of those contexts and variations. The diagnostic manual provides tools for self-diagnosis and emerges as an apparatus for taking care for the self, to fashion the self as always already a medical subject. This responsible sexual subject materialises from the therapeutic encounter and carries on the labour of analysis on the self. The deployment of sexuality in the twenty-first century is now laden with the language of ethics and a focus on the need for balance in sexual appetite, as exemplified by the DSM. Concurrently, the DSM requires that the physician oversee the instrument of classification. The application of the technique of classification in the manual involves the dual requirements of self-diagnosis and professional expertise.

This chapter has demonstrated how appetite became central to conceptualisations of sexual disorders in the DSM in the late twentieth and twenty-first centuries. However, alongside revisions of the diagnostic manual, this period witnessed the intensification in psychiatry of the development and prescription of pharmaceuticals for sexual imbalance. In Chap. 6, the book explores the emergence of Addyi (flibanserin) as a case study of how this technique produces a particular subject of pharmaceutical knowledge. The chapter considers the significance of the act of pharmaceutical ingestion on the embodied subjectivity of the consumer and the chemical constitution of the human body. The use of Addyi to manage sexual imbalance in combination with the diagnostic manual converges in the emergence of a socio-technical and knowledge-gathering subject. This subject is armed with tools to monitor the self and gather knowledge of her sexual appetite, a process that affirms intimacy. Operating alongside

[73] See the special issue on "Hypersexual Disorder" of *Sexual Addiction & Compulsivity: The Journal of Prevention and Treatment* 20, no. 1–2 (2013): 1–153. For a compelling critique of sexuality and "addiction discourse," see Helen Keane, "Disorders of Desire: Addiction and Problems of Intimacy," *Journal of Medical Humanities* 25, no. 3 (2004): 189–204.

[74] APA, DSM-5, 423.

the diagnostic manual, Addyi acts on the chemical composition of the brain and embeds techniques of management of sexual appetite within the body itself. Chapter 6 thus explores how pharmaceutical intimacy as a technique of self-management represents sexual subjectivity mediated by written text, spoken word and chemical interactions.

Bibliography

American Psychiatric Association. 1952. *Diagnostic and Statistical Manual of Mental Disorders*. 1st ed. Washington, DC: American Psychiatric Association.

———. 1968. *Diagnostic and Statistical Manual of Mental Disorders*. 2nd ed. Washington, DC: American Psychiatric Association.

———. 1980. *Diagnostic and Statistical Manual of Mental Disorders*. 3rd ed. Washington, DC: American Psychiatric Association.

———. 1987. *Diagnostic and Statistical Manual of Mental Disorders*. 3rd ed. Revised. Washington, DC: American Psychiatric Association.

———. 1994. *Diagnostic and Statistical Manual of Mental Disorders*. 4th ed. Washington, DC: American Psychiatric Association.

———. 2013. *Diagnostic and Statistical Manual of Mental Disorders*. 5th ed. Washington, DC: American Psychiatric Association.

Bayer, Ronald. 1987 [1981]. *Homosexuality and American Psychiatry: The Politics of Diagnosis*. Princeton: Princeton University Press.

Bowker, Geoffrey C., and Susan Leigh Star. 1999. *Sorting Things Out: Classification and Its Consequences*. Cambridge, MA: MIT Press.

Bronski, Michael. 2011. *A Queer History of the United States: ReVisioning American History*. Boston: Beacon Press.

Brookey, Robert Alan. 2002. *Reinventing the Male Homosexual: The Rhetoric and Power of the Gay Gene*. Bloomington, IN: Bloomington University Press.

Clayton, Anita H., R.T. Segraves, D. Bakish, D. Goldmeier, J. Tignol, R.H. van Lunsen, R.E. Nappi, et al. 2010. Cutoff Score for the Inventory-Female for Diagnosis of Hypoactive Sexual Desire Disorder. *Journal of Women's Health* 19 (12): 2191–2195.

Cooper, Rachel. 2004. What's Wrong with the DSM? *History of Psychiatry* 15 (1): 5–25.

———. 2007. *Psychiatry and the Philosophy of Science*. Montreal and Kingston: McGill-Queen's University Press.

Corona, G., E.A. Jannini, and M. Maggi. 2006. Review: Inventories for Male and Female Sexual Dysfunctions. *International Journal of Impotence Research* 18 (3): 236–250.

Decker, Hannah S. 2013. *The Making of the DSM-III: A Diagnostic Manual's Conquest of American Psychiatry*. Oxford: Oxford University Press.

D'Emilio, John, and Estelle B. Freedman. 2012 [1988]. *Intimate Matters: A History of Sexuality in America*, 3rd ed. New York: Harper & Row.

Downing, Lisa. 2015. Heteronormativity and Repronormativity in Sexological 'Perversion Theory' and the DSM-5's 'Paraphilic Disorder. *Archives of Sexual Behavior* 44 (5): 1139–1145.

Drescher, Jack. 2010. Queer Diagnoses: Parallels and Contrasts in the History of Homosexuality, Gender Variance, and the *Diagnostic and Statistical Manual. Archives of Sexual Behavior* 39 (2): 427–460.

———. 2012. The Removal of Homosexuality from the DSM: Its Impact on Today's Marriage Equality Debate. *Journal of Gay & Lesbian Mental Health* 16 (2): 124–135.

Foucault, Michel. 1978. *The Will to Knowledge: The History of Sexuality, Volume 1*. Trans. Robert Hurley. London: Penguin Books.

———. 1985. *The Use of Pleasure: History of Sexuality, Volume 2*. Trans. Robert Hurley. New York: Vintage Books.

———. 1986. *The Care of the Self: The History of Sexuality, Volume 3*. Trans. Robert Hurley. New York: Vintage Books.

———. 1988. Technologies of the Self. In *Technologies of the Self: A Seminar with Michel Foucault*, eds. Luther H. Martin, Huck Gutman, and Patrick H. Hutton, 16–49. Amherst: University of Massachusetts Press.

———. 1997. Subjectivity and Truth. In *Ethics: Essential Works of Foucault, 1954–1984, Volume 1*, ed. Paul Rabinow, 87–92. London: Penguin Books.

Frances, Allen. 2013. *Saving Normal: An Insider's Revolt Against Out-of-Control Psychiatric Diagnosis, DSM-5, Big Pharma, and the Medicalization of Ordinary Life*. New York: William Morrow.

Giami, Alain. 2015. Between DSM and ICD: Paraphilias and the Transformation of Sexual Norms. *Archives of Sexual Behavior* 44 (5): 1127–1138.

Harsh, Veronica, Elizabeth Lloyd McGarvey, and Anita H. Clayton. 2008. Physician Attitudes Regarding Hypoactive Sexual Desire Disorder in a Primary Care Clinic. *Journal of Sexual Medicine* 5: 640–645.

Houts, Arthur C. 2000. Fifty Years of Psychiatric Nomenclature: Reflections on the 1943 War Department Technical Bulletin, Medical 203. *Journal of Clinical Psychology* 56 (7): 935–967.

Irvine, Janice. 2005. *Disorders of Desire: Sex and Gender in Modern American Sexology*. Philadelphia: Temple University Press.

Kaplan, Helen Singer. 1979. *Disorders of Desire and Other Concepts and Techniques in Sex Therapy*. New York: Simon & Schuster.

Kaplan, Arline. 2009. DSM-V Controversies. *Psychiatric Times* 26 (1): 5–10.

Keane, Helen. 2004. Disorders of Desire: Addiction and Problems of Intimacy. *Journal of Medical Humanities* 25 (3): 189–204.

Kirk, Stuart A., and Herb Kutchins. 1992. *The Selling of the DSM: The Rhetoric of Science in Psychiatry*. New Brunswick: Transaction.
Klerman, Gerald L., George E. Vaillant, Robert L. Spitzer, and Robert Michels. 1984. A Debate on DSM-III: The Advantages of DSM-III. *American Journal of Psychiatry* 141 (4): 539–553.
LeVay, Simon. 2010. *Gay, Straight and the Reason Why*. Oxford: Oxford University Press.
Lief, Harold. 1977. Inhibited Sexual Desire. *Medical Aspects of Human Sexuality* 7: 94–95.
Masters, William H., and Virginia E. Johnson. 1970. *Human Sexual Inadequacy*. Boston: Little, Brown & Company.
———. 1994. *Heterosexuality*. New York: Harper Collins.
Maxmen, Jerrold S. 1985. *The New Psychiatry: How Modern Psychiatrists Think about Their Patients, Theories, Diagnoses, Drugs, Psychotherapies, Power, Training, Families and Private Lives*. New York: William Morrow & Company.
Mayes, Rick, and Allan V. Horwitz. 2005. DSM-III and the Revolution in the Classification of Mental Illness. *Journal of the History of Behavioral Sciences* 41 (3): 249–275.
Mol, Annemarie. 2002. *The Body Multiple: Ontology in Medical Practice*. Durham and London: Duke University Press.
Paras, Eric. 2006. *Foucault 2.0: Beyond Power and Knowledge*. New York: Other Press.
Rose, Nikolas. 1989. *Governing the Soul: The Shaping of the Private Self*. London and New York: Routledge.
———. 1998. *Inventing Our Selves: Psychology, Power, and Personhood*. Cambridge: Cambridge University Press.
———. 2007. *The Politics of Life Itself: Biomedicine, Power, Subjectivity in the Twenty-First Century*. Princeton: Princeton University Press.
———. 2019. *Our Psychiatric Future: The Politics of Mental Health*. Cambridge: Polity.
Sabshin, Melvin. 1990. Turning Points in Twentieth Century Psychiatry. *American Journal of Psychiatry* 147 (10): 1267–1274.
Schover, Leslie R., Jerry M. Friedman, Stephen J. Weiler, Julia R. Heiman, and Joseph LoPiccolo. 1982. Multiaxial Problem-Oriented System for Sexual Dysfunctions. *Archives of General Psychiatry* 39: 614–619.
Segal, Lynne. 1994. *Straight Sex: Rethinking the Politics of Pleasure*. Berkeley: University of California Press.
Shorter, Edward. 2005. *Historical Dictionary of Psychiatry*. Oxford: Oxford University Press.
Singh, Ilina, and Nikolas Rose. 2009. Biomarkers in Psychiatry. *Nature* 460: 202–207.

Spitzer, Robert L., and Janet B.W. Williams. 1983. The Revision of the DSM-III. *Psychiatric Annals* 13 (10): 808–811.

Tiefer, Leonore. 2004. *Sex Is Not a Natural Act and Other Essays.* New York: Westview Press.

Weeks, Jeffrey. 2015. Beyond the Categories. *Archives of Sexual Behavior* 44 (5): 1091–1097.

CHAPTER 6

The Sexual Pharmacy

In 1998, the Food and Drug Administration (FDA) in the United States approved the first oral drug designed for the treatment of erectile dysfunction, sildenafil citrate (Viagra). Difficulties in achieving satisfactory and functional erections had been a longstanding issue for men and their partners.[1] Prior to the marketing of Viagra, the penis was the direct site of medical intervention through penile pumps, prostheses, injections at the base of the organ and the implantation of silicon rods. These apparatuses optimised the penis for sexual activity through a mechanical, hydraulic reaction. The pharmacological cure that naturalised the process of preparing the penis for sex was heralded as profoundly changing sexual mores as it promised control and choice.[2] Since 1998, multinational pharmaceutical companies including Pfizer and Boehringer Ingelheim have sought to develop an equivalent tablet for women. In August 2015, Sprout Pharmaceuticals obtained approval from the FDA for flibanserin, marketed

[1] As demonstrated in Angus McLaren, *Impotence: A Cultural History* (Chicago and London: University of Chicago Press, 2007).

[2] See, for example, Jack Hitt, "The Second Sexual Revolution," *The New York Times*, February 20, 2000, accessed January 19, 2019. http://www.nytimes.com/2000/02/20/magazine/the-second-sexual-revolution.html?pagewanted=all, and Steven Lamm, and Gerald Secor Couzens, *The Virility Solution: Everything You Need to Know about Viagra, the Potency Pill that Can Restore and Enhance Male Sexuality* (New York: Fireside Books, 1998). It is worth noting that drugs for the management of male sexual appetite developed rapidly after Viagra. They include Cialis, Staxyn, Stendra, Edex and Levitra, though there are differences between how they work.

J. Flore, *A Genealogy of Appetite in the Sexual Sciences*,
https://doi.org/10.1007/978-3-030-39423-3_6

as Addyi, a drug for premenopausal women distressed by a lack of sexual desire.[3] Unlike Viagra, which acts to produce hardness, Sprout Pharmaceuticals claims that Addyi targets the female brain by increasing the levels of the neurotransmitters dopamine and noradrenaline and lowering levels of serotonin.

The pharmaceutical tablet changes the composition of the body chemically, by embedding techniques for managing sexual appetite in the body itself. Addyi, which was originally trialled as an antidepressant, is believed to act on brain chemicals connected to mood and appetite. In such discourses, neuroreceptors of the female brain are conceptualised as plastic, malleable and alterable. The problem of low sexual appetite in women is thus located firmly in the brain; it is a problem of balance within neuroreceptors, and of *synchronicity* between sexual organs and cerebral neurotransmitters. However, in a discussion on depression drugs, Wilson argues that the gut and other biological systems are implicated in the functioning of pharmaceuticals designed to affect individuals' moods.[4] Thus, the chemical composition of Addyi is both intended for the brain and designed to exert influence on sexual appetite, while becoming entangled in embodiment as it travels through the body.

This chapter examines the pharmaceutical tablet as a technique for the management of sexual appetite in the twenty-first century. It explores the emergence of Addyi as a case study of how this technique produces a particular subject of pharmaceutical knowledge. The chapter considers the significance of the act of pharmaceutical ingestion on the embodied subjectivity of the consumer and the chemical constitution of the human body. The use of Addyi to manage sexual imbalance in combination with the tools of the diagnostic manual, discussed in Chap. 5, converges in the emergence of a socio-technical and knowledge-gathering subject. This subject is armed with techniques to monitor the self and gather knowledge

[3] A couple of days later, Valeant Pharmaceuticals International announced that it had acquired Sprout for $1 billion. Another drug Vyleesi (bremelanotide) was approved by the FDA in 2019. Much like Addyi, Vyleesi claims to target neurological pathways by increasing levels of dopamine in the brain. While there are differences between how the two drugs work—Vyleesi, for example, is used through subcutaneous injection 45 minutes before sexual activity—the way they produce understandings of sexual appetite in women is similar: it can be addressed through "working on" the brain. See Food and Drug Administration. "Drug Trials Snapshot: Vyleesi." 2019, accessed July 19, 2019. https://www.fda.gov/drugs/drug-safety-and-availability/drug-trials-snapshots-vyleesi.

[4] Elizabeth A. Wilson, *Gut Feminism* (Durham & London: Duke University Press, 2015), 100.

of her sexual imbalance, a process that affirms intimacy. Indeed, the subject who swallows the pill is a fundamentally social one, that is to say, one who desires intimate contact with others.

Prescribing Sex

In Western liberal societies, affects, moods and desires are routinely managed, tinkered with or altogether transformed through the ingestion of pharmaceutical tablets. The birth control pill, aspirin, penicillin, Viagra and mood stabilisers shape and mould norms, relations and practices. Medicine developed tools, diagnoses and techniques for managing sexual appetite across the nineteenth to twenty-first centuries, which in turn reshaped what it meant to be "human." The rapid expansion of machines of diagnosis has been compounded with the infiltration of pharmaceutical products in everyday life. Nikolas Rose describes twenty-first-century life as "psychopharmacological" to account for the emergence and prevalence of medicinal drugs in contemporary life.[5] While the use of such drugs to treat psychiatric ailments began around the 1950s, Rose identifies a rapid expansion in the sale and prescription of psychopharmacological products in the twenty-first century resulting in a transformation in conceptions of life and personhood.

The goal of psychopharmacological drugs, Rose asserts, is now less focused on correcting deviance and more concerned with the management of everyday life. The discourses and strategies employed in such societies have as their aim the transformation of bodies, the improvement of life and "an ethic of self-control, lifestyle promotion, and self-realization."[6] The drugs are presented as tools for a self that is not thoroughly changed, but rather enhanced to a "better" version, a subject who is more able to engage with the various demands of everyday life. The management of everyday life through the psychopharmacological tablet functions with the input of the patient. Through the ingestion of the psychopharmacological product, the subject participates in a practice of self-scrutiny where moods, emotion and cognition are observed, and patterns are mapped. Paul Preciado argues that a similar discourse of "memory and time, responsibility

[5] Nikolas Rose, *The Politics of Life Itself: Biomedicine, Power, and Subjectivity in the Twenty-First Century* (Princeton: Princeton University Press, 2007), 209.

[6] Ibid., 212. See also Nikolas Rose, "Neurochemical Selves," *Society* 41, no. 1 (2003): 46–59.

and trust" circulates in the packaging and selling of the birth control pill.[7] The subject is involved in diagnosis and treatment, and must also be responsible for managing and monitoring her or his capacities.[8] This form of biopower is centred on individual responsibility and quality of life. It is, as Rose writes, "the government of *life*."[9] It is thus important to approach the chemical tablet as a convergence of different objectives and different actors within the discourses of consumption, health and sexuality.

The language of medicalisation has been deployed in critiques of the development and approval of drugs for managing sexual appetite to imply "passivity on the part of the medicalised,"[10] suggesting a subject on whom medical expertise is imposed and who has little understanding of her/his condition, or their treatment. However, such critiques misconceive how narratives of diagnosis and treatment are regularly appropriated and reshaped by individuals.[11] Conceiving of diagnoses and drugs as "corporate sponsored creation"[12] risks positioning the subject as subordinate to the will of the expert. In the twenty-first century, the subject who ingests psychopharmacological products needs to be conversant with discourses of diagnoses and treatment that tend to reproduce ideas of resilience, self-fashioning and responsibility. The consumer emerges as a willing participant who actively accommodates the drug into an everyday ritual of chemical absorption. Diseases and pharmaceutical products are not developed in isolation from potential consumers, and medical personnel do not coerce individuals. The rhetoric of health and illness deployed by medical science might provide a vocabulary to express a set of circumstances, but this evokes not a "false" condition or remedy; rather it constitutes "the creation of delicate affiliations between subjective hopes and dissatisfactions and the alleged capacities of the drug."[13] A medicalisation of sexual appetite emerges within medical and popular discourses of what counts as "enough" sex and the understanding of sex as a "healthy" and necessary

[7] Beatriz (Paul) Preciado, *Testo Junkie: Sex, Drugs, and Biopolitics in the Pharmacopornographic Era*, trans. Bruce Benderson (New York: The Feminist Press, 2013), 198.

[8] Rose, *The Politics of Life Itself*, 223.

[9] Ibid., 70 (emphasis original).

[10] Nikolas Rose, "Beyond Medicalisation," *The Lancet* 369 (2007): 702.

[11] Examples include depression, myalgic encephalomyelitis (Chronic Fatigue Syndrome), persistent genital arousal disorder, sexual pain and bipolar disorder.

[12] Roy Moynihan, "The Making of a Disease: Female Sexual Dysfunction," *British Medical Journal* 326, no. 7379 (2003): 45.

[13] Rose, "Beyond Medicalisation," 702.

aspect of human life. This is a subject who desires and demands enjoyable embodied sexual experiences, which produce a complex intermeshing of consumerism, health and sexual pleasure.[14]

The psychopharmacological configuration of the body-subject is located at the nexus of knowledge-gathering, technological inventions and the ingestion of medicinal drugs. The body that consumes drugs for sexual enhancement shapes embodied subjectivity through the technological arrangement of chemical reactions. The sexual pharmaceutical transforms the constitution of the body. It becomes infused with chemicals, but also with the technology of the tablet, which mediate reactions, affects and intimacy. The ability of the body to respond "naturally" to a sexual interaction through the hardness, engorgement and wetness of genital organs is both optimised and moderated through the medical capsule. However, instead of viewing this body through the prism of a natural/unnatural (or chemical) dualism, the sexual pharmaceutical obscures such distinctions. In her seminal "Cyborg Manifesto," Donna Haraway suggests that technology and science are inextricable from what it means to be human. Indeed, Haraway argues that we are living in an age of intensified machine-body relations where the cyborg is a "cybernetic organism, a hybrid of machine and organism."[15] The machine is an integral aspect of human embodiment, thus offering "a way out of the maze of dualisms in which we have explained our bodies."[16]

The boundaries between human and machine, or human and information, have become increasingly blurred in the twenty-first century. Preciado deploys the term "pharmacopornographic regime" to refer to processes of "biomolecular (pharmaco) and semiotic-technical (pornographic) government of sexual subjectivity."[17] This regime is characterised by "medicolegal surveillance and mediatic spectacularization [and] intensified … by digital and data-processing techniques and communication networks."[18] Human bodies are generative of data used to assess and measure various ailments and performance across time and space. Sexual performance and

[14] See Kane Race, *Pleasure Consuming Medicine: The Queer Politics of Drugs* (Durham and London: Duke University Press, 2009).

[15] Donna J. Haraway, *Simians, Cyborgs, and Women: The Reinvention of Nature* (New York: Routledge, 1991), 272.

[16] Ibid., 325–327.

[17] Preciado, *Testo Junkie*, 33–34.

[18] Ibid., 76.

sexual enjoyment are extensively measured through techno-scientific tools.[19] Knowledge harvested from technological equipment is translated into tools that are increasingly integrated within human bodies to control, enhance and maximise performance. As Preciado argues, "[t]he pharmacopornographic body is not a passive living matter but a techno-organic interface, a technoliving system segmented and territorialized by different (textual, data-processing, biochemical) political technologies."[20] The sexual pharmaceuticals create a cybernetic entity, a subjectivity that is both transformed through chemical formulae and techno-scientific apparatuses, as the subject is also framed as better able to respond to certain situations through chemical ingestion. In other words, the integration of a pharmaceutical diet equips the subject to navigate social life while transforming the chemical constitution of the body.

Drawing on the writings of Haraway, Annie Potts uses the concept of "Viagra cyborg" to elucidate the kind of embodied subjectivity that emerges from the consumption of Viagra.[21] For Potts, the Viagra cyborg "does not represent a human–machine recomposition; he is an entity created through a human–drug relationship."[22] The Viagra cyborg evokes both a reparation of sexual appetite and a heightened, almost exaggerated, notion of potency and performance. I would argue, however, that the kind of cyborg subjectivity that emerges from the ingestion of sexual pharmaceuticals does not completely evacuate the machine from its constitution. Rather, this cyborg subjectivity encompasses machines, technology and chemical drugs. The subject who ingests the sexual pharmaceutical represents a confluence, and a production, of different forms of knowledge that exceed the drug itself. The human-drug relation is only one part of the consumption of sexual pharmaceuticals. While the metaphor of machinery, pipes and hydraulics is extensively used in the marketing of drugs such as Viagra, the narratives of effectiveness also reaffirm natural and biological sexual experiences. The pharmaceutical drug aims to facilitate what is already occurring in chemical reactions in the body. However, at the same time, the body that ingests this substance represents an enhanced version

[19] Donna J. Drucker, *The Machines of Sex Research: The Machines of Sex: Research Technology and the Politics of Identity, 1945–1985* (Dordrecht: Springer, 2014).

[20] Preciado, *Testo Junkie*, 114.

[21] Annie Potts, "Cyborg Masculinity in the Viagra Era," *Sexualities, Evolution and Gender* 7, no. 1 (2005): 3–16.

[22] Ibid., 4.

of natural, biological sexual appetite. In fact, the drug manufacturers of Addyi are careful to distance themselves from the language of performance-enhancing drugs, which can imply an "unnatural" modification of the body.[23] Sexual pharmaceuticals differ from performance-enhancing drugs insofar as they are restorative agents that reinstate desire where it has disappeared, facilitate natural bodily responses and maximise capacity and functionality, but at the same time purport to return sexual appetite to its "natural" and optimal state.

The body in twenty-first-century medical knowledge thus occupies an ambiguous position. It is both treated as possessing biological capacities that can be enhanced and as particularly amenable to being manipulated through chemical reactions. It is both a self-contained entity with boundaries that cannot be exceeded and an object whose fleshy interior can be examined on a molecular level. Even as inventories are developed, taxonomy is updated and chemical tablets are manufactured, the body is continually believed to be ahistorical, isolated from its sociocultural contexts and examinable in the decontaminated, sterile space of the research laboratory.

In the production of sexual pharmaceutical tablets, an interpretation of what a body does and what it can do occurs through a confluence of meanings and signification that are not exclusively medical. As Thea Cacchioni observes, sexual medicine has often employed the rhetoric of psychology to cover the socio-political, cultural and political bases of sexual problems. This is partly due to the struggle of claiming the territory as a scientific field, and obtaining industry funding.[24] Nonetheless, the notion of so-called psychogenic aspects of sexual problems has been criticised by some feminist sexologists as minimising socio-political, cultural and economic factors that impact on sexuality.[25] The gendering of sexual appetite in psychiatric nosology and in the production of sexual pharmaceuticals

[23] Under "Indication" on Addyi's website: "Addyi is not … to improve sexual performance," accessed January 19, 2019. https://www.addyi.com/. This narrative is common to Vyleesi, whose label also states that it is not indicated to enhance sexual performance. See Food and Drug Administration. "Vyleesi Label." 2019, accessed July 19, 2019. https://www.accessdata.fda.gov/drugsatfda_docs/label/2019/210557s000lbl.pdf.

[24] Thea Cacchioni, *Big Pharma, Women, and the Labour of Love* (Toronto: University of Toronto Press, 2015), 38–43.

[25] See Leonore Tiefer, *Sex Is Not a Natural Act and Other Essays* (New York: Westview Press, 2004); "Arriving at a 'New View' of Women's Sexual Problems: Background, Theory and Activism," *Women & Therapy* 24, no. 1–2 (2002): 63–98; and "The Viagra Phenomenon," *Sexualities* 9, no, 3 (2006): 273–294.

reflects this tension. For Marshall, "*Gender* is never absent for women in the way it can be rendered invisible for men"[26] in medical discourse. For women, issues such as intimacy and body image "are frequently cited as 'confounding' variables in sorting out the physiology of female sexual response."[27]

The production of a pharmaceutical tablet to treat a lack of sexual appetite in women reveals the gendering of sexual medicine. Viagra, for example, assumes that male desire is already present, and the pill merely acts as a conduit to manifest desire through hardness. On the other hand, the ingestion of Addyi reifies the prime location of female sexual appetite as the brain. This represents an important turn in the development of drugs for the management of sexual appetite. Not only is it claimed as the first drug that treats lack of sexual appetite in women, the drug also represents the heightened gendering of medical knowledge by situating the problem of appetite in the cerebral cortex. Similar to the "accidental" discovery of sildenafil citrate (Viagra), which was originally tested as medication for high blood pressure, flibanserin (Addyi) was found to have an effect on sexual appetite when tested on women "whose depressive symptoms included decreased sexual desire at baseline."[28] These discoveries exemplify Wilson's point that "biological data often look much more like they are describing networks of affinity."[29] While Addyi is marketed as targeting neurotransmitters in the brain, it also reveals the co-constitution of organs and affects; a drug targeting moods is found to exert influence on sexual interest. Further testing on female rats in 2013, which was supported by an unrestricted grant from Boehringer Ingelheim, revealed that it was possible that flibanserin could cause an increase in "female sexual motivation."[30] As noted in the previous chapter, in 2013, the DSM-5 positioned "interest" in sex as a crucial factor in women's sexual function—a criterion that

[26] Barbara L. Marshall, "'Hard Science': Gendered Constructions of Sexual Dysfunction in the 'Viagra Age,'" *Sexualities* 5, no. 2 (2002): 141 (emphasis original).

[27] Ibid., 141.

[28] A.H. Clayton, L. Dennerstein, R. Pyke, and M. Sand, "Flibanserin: A Potential Treatment for Hypoactive Sexual Desire Disorder in Premenopausal Women," *Women's Health* 6, no. 5 (2010): 639–653.

[29] Wilson, *Gut Feminism*, 35.

[30] Helene Gelez, Pierre Clement, Sandrine Compagnie, Diane Gorny, Miguel Laurin, Kelly Allers, Bernd Sommer, and Francois Giuliano, "Brain Neuronal Activation Induced by Flibanserin Treatment in Female Rats," *Psychopharmacology* 230 (2013): 639–652.

does not exist for men.[31] Interest aims to cover how women sense and experience sexual attraction in their minds.[32] In a cyborg body where the line between natural and artificial boundaries is blurred, the subject must also be attentive to the constitution of their minds, and this, it appears, is mainly a problem for women in sexual medicine.

Twenty-first-century inventories of psychiatric disorders combine the biological manifestation of "arousal," for example, vaginal congestion and humidification, with the problem of affective and emotional inclinations. This turn in nomenclature represents both a reification of female issues of synchronicity between mind and body, and an inclusion of the social world of subjects. While women are not always attuned to the lubrication in their genitals, vaginal congestion and clitoral erection are nonetheless taken as impartial and objective markers of sexual arousal in women. Scientific approaches to female sexuality inscribe women as needing synchronicity between the mind and the body. While Viagra involves one tablet taken before sexual activity, Addyi, like the birth control pill, necessitates a daily intake. Sexual appetite, through the ingestion of Addyi, is situated in the brain, and requires a daily chemical input in order to achieve balance. It also requires the female subject to take care and monitor the constitution of her brain. In other words, women must be attuned to changes at the level of the brain in terms of their ability to recognise sexual events and their interest in them. Though Addyi might be approached as an "old antidepressant" that has been repackaged and branded as a sexual pharmaceutical,[33] the drug nonetheless represents a remarkable turn in the

[31] While "interest" is not positioned as an issue for men in the DSM, the DSM-5 introduced in 2013 the category of Male Hypoactive Desire Disorder. See American Psychiatric Association, *Diagnostic and Statistical Manual of Mental Disorders*, 5th edition (Washington, DC: American Psychiatric Association, 2013), 440–443.

[32] "Interest" has also been employed to suggest that women are more likely to experience desire *in response to* an initiating partner/image. See Anthony F. Bogaert and Lori A. Brotto, "Object of Desire Self-Consciousness Theory," *Journal of Sex and Marital Therapy* 40, no. 4 (2014): 323–338.

[33] Jayne Lucke, "A Sexually Satisfying Event for Women, or Just a New Identity for an Old Antidepressant," *The Conversation*, June 6, 2015, accessed January 19, 2019. https://theconversation.com/a-sexually-satisfying-event-for-women-or-just-a-new-identity-for-an-old-antidepressant-42734. Addyi, under the name BIMT 17, was originally trialled as a potential antidepressant. See F Borsini, E Giraldo, E Monferini, G Antonini, M Parenti, G Bietti, and A Donnetti, "BIMT 17, a 5-HT2A receptor antagonist and 5-HT1A receptor full agonist in rat cerebral cortex," *Naunyn Schmiedebergs Arch. Pharmacol* 352, no. 3 (1995): 276–282.

production of drugs for sexual enhancement. It reveals, as Wilson writes, that drugs are not "autocratic agents that operate unilaterally on body and mind; rather, they are substances that find their pharmaceutical efficacy by being trafficked, transformed, and broken down."[34]

When to take Addyi (nightly before bed), what not to consume (alcohol), duration of treatment (three to six months) and potential side effects (e.g. dizziness, nausea and fainting), all form part of the action of the pharmaceutical drug. This involves not only the brain but also blood pressure, the gut and the neurological mechanisms involved in regulating sleep. All these functions are not at the periphery of the drug, they are "a decisive part of that drug's psychological punch."[35] Addyi thus might be intended for the brain and its rationale might locate sexuality in the brain, yet its complex functioning within the body reveals its entanglements.

Addyi demonstrates how sexual balance becomes centralised and localised in both the brain and social practice. It functions to increase levels of the neurotransmitter dopamine and noradrenaline, and to lower levels of serotonin. It acts as a messenger of equilibrium between excitation and inhibition. Inhibition is not so much concerned with the complete sublimation of urges and needs; instead, it involves a management of sensations and affective connections to attain balance in social life. The notion of subjectivity that emerges from this emphasis on the brain and balance is mediated through chemical reactions. The Executive Summary submitted by Sprout Pharmaceuticals to the FDA in 2015 notes, underneath a drawing of a human brain, that flibanserin restores "appropriate balance of excitatory and inhibitory activity of reward structures to the prefrontal cortex. The net result is a greater ability for premenopausal women with HSDD [Hypoactive Sexual Desire Disorder] to feel sexual desire when appropriate."[36] Through a drug such as Addyi, the patient is invited to understand her sexual interest in terms of chemical imbalance located in the brain. Balance is to be achieved by taking care of the self at a molecular level, through an understanding of one's neurotransmitters.[37] However, as Wilson demonstrates, drugs intended for the brain do not solely affect the brain. Organs are "always already coevolved and coentangled."[38] Addyi

[34] Wilson, *Gut Feminism*, 102.

[35] Ibid.

[36] Sprout Pharmaceuticals, *Flibanserin for the Treatment of Hypoactive Sexual Desire Disorder in Premenopausal Women NDA 022526*, Advisory Briefing Document, 2015, 15.

[37] See Rose, *The Politics of Life Itself*, 143.

[38] Wilson, *Gut Feminism*, 66.

passes through the digestive system before reaching the brain; its effects encompass not only "sexualising" the brain but also producing an ensemble of effects at different locations in the body.

In the twenty-first century, humans are encouraged to be the agents of their own regulation through chemical drugs. The regulation occurs in areas such as sexuality, mood and nutrition. Central to a technology of biopower then are the notions of balance and regulation. For Foucault, biopower is "continuous, scientific"[39] and medicine has "both disciplinary effects and regulatory effects."[40] The pharmaceutical tablet is a technology of biopower insofar as it mediates capitalist reproduction and the futurity of the heterosexual couple. Medical knowledge on sexuality, as a technique of biopower, is not limited to the documentation of the fertility of the population, one of the paradigmatic procedures identified by Foucault. Through the ingestion of the drug to manage sexual appetite, this technique of biopower becomes much more intimate and indistinguishable from the subject who swallows the pill. For Preciado, biopower through the tablet now "dwells at home, sleeps with us, inhabits within."[41] There is a compounding of medical knowledge, manuals and inventories with the mundane, routine ingestion of the drug. Preciado, in an analysis of the birth control pill, notes that this is not a form of power that is imposed or that invades from outside the body; "it is the body desiring power, seeking to swallow it, eat it, administer it."[42] Drugs such as Addyi and Viagra are not simply devices of control, regulation and discipline; they are implicated in the formation of subjectivity. Hence, in its purported effects, Addyi partakes in a reconfiguration of embodied subjectivity on a molecular, more intimate level where the brain (allegedly) houses sexuality and becomes the locus of intervention through chemicals.

Brain circuitry emerges as an entity that can be governed and reshaped in the case study of Addyi. It represents an opportunity to enhance sexuality and life itself. Through an interaction between commercial interests, therapeutic demand and chemicals, the brain is perceived as possessing chemical messengers of sexual appetite, and promises that through a cognitive enhancement, one's affects and senses will be better equipped to be

[39] Michel Foucault, *Society Must Be Defended: Lectures at the Collège de France, 1975–76*, trans. David Macey (London: Penguin Books, 2003), 247.

[40] Ibid., 252.

[41] Preciado, *Testo Junkie*, 207.

[42] Ibid., 208.

sensitive to a sexual event and act on it. In their extensive study of the brain sciences, Nikolas Rose and Joelle M. Abi-Rached observe that "the human brain has come to be anatomized at a molecular level, understood as … exquisitely adapted to human interaction and sociality, and open to investigation at both the molecular and systemic level in a range of experimental setups."[43] Through its regular, routine ingestion, and its promise of acting on chemicals and on relationships, the tablet encourages the subject to participate in a socially intimate practice mediated by pharmaceuticals.

Pharmaceutical Intimacy

In *Neuro*, Rose and Abi-Rached demonstrate that the emergence of techniques for the management of brains has incorporated a concern with sociality. In the hypothesis of the "social brain," capacities for sociality and connection are "neurally located" in certain regions of the brain.[44] The human brain then becomes a product of one's social and cultural environments, while at the same time being amenable to change as our environments themselves change. The importance of neurochemicals in the history of sexuality is nothing new, as empirical scientific research continues to situate sexual desire and pleasure in cerebral circuitry.[45] But just as brains are increasingly conceived as social organs, or as organs responsive to sociality, it now appears that "capacities that are crucial to society are a matter of brains. … And we have a social brain in that this organ is now construed as malleable, open to, and shaped by, social interactions—shaping sociality as it is itself reshaped by it."[46] The pharmaceutical tablet that is ingested for the enhancement of sexual appetite, or the one that is swallowed to facilitate a sexual experience, reiterates the norm that sexuality

[43] Nikolas Rose and Joelle M. Abi-Rached, *Neuro: The New Brain Sciences and the Management of the Mind* (Princeton and New York: Princeton University Press, 2013), 9.

[44] Ibid., 143.

[45] See, for example, J.R. Georgiadis and M.L. Kringelbach, "The Human Sexual Response Cycle: Brain Imaging Evidence Linking Sex to Other Pleasures," *Progress in Neurobiology* 98 (2012): 49–81, Serge Stoléru, Véronique Fonteille, Christel Cornélis, Christian Joyal, and Virginie Moulier, "Functional Neuroimaging Studies of Sexual Arousal and Orgasm in Healthy Men and Women: A Review and Meta-Analysis," *Neuroscience and Biobehavioral Reviews* 36 (2012): 1481–1509, and David L. Rowland and Ion G. Motofei, "The Mind and Sexuality: Introduction to a Psychophysiological Perspective," *Journal of Mind and Medical Sciences* 2, no. 1 (2015): 1–8.

[46] Rose and Abi-Rached, *Neuro*, 163.

and intimacy are necessary aspects of life. Addyi, in particular, locates durable bonds as an imperative of normal cognitive functioning. As its website proclaims, above the picture of a heterosexual couple in bed, "Your brain may be working against you when it comes to sex."[47] By acting to increase sexual interest in the brain, in other words, the pharmaceutical pill also organises the material and social world of the subject by ensuring that the sexual event is identified and, ideally, acted upon.

The pharmaceutical tablet swallowed to manage sexual appetite represents a technique of intimate sociality. And this form of intimacy does not evade normative conceptualisation. Early advertising material for Viagra featured middle-aged heterosexual couples attesting to durable, monogamous bonds. One ad for instance included the line "let the dance begin,"[48] suggesting that Viagra functions as a technique to strengthen committed couples through encouraging sexual intimacy. The hearing for Addyi at the FDA reveals a similar narrative of enduring heteronormative monogamy.[49] The loss of sexual appetite was framed as affecting strong intimate relations with an ongoing partner, sometimes following pregnancy. The implication being that the ingestion of the drug would strengthen those relationships that were already solid, but are now strained by low sexual appetite. This narrative serves to ossify the "medical" aspect of the drug and minimise the possibility of the drug to be used for recreational purposes such as casual sex. The daily intake of the drug then not only functions to "train" the brain's neurotransmitters and neuroreceptors, but also serves to act on the social world by turning committed heterosexual couples into resilient subjects—resilience in terms of both the relationship and the longevity of their socio-sexual life.

Certainly, the sexual pharmaceutical may be especially targeted towards long-term, monogamous heteronormative couples, but drugs always embody a potential for subversion and "pure" pleasure, as the story of Viagra has demonstrated. Towards the end of the twentieth century, the "lifestyle drug"—which includes Viagra—emerged as a particular kind of pharmaceutical object. The lifestyle drug is often defined in consequential

[47] Addyi, "Addyi (flibanserin)," 2019, accessed January 19, 2019. https://addyi.com/.

[48] Meika Loe, *The Rise of Viagra: How the Little Blue Pill Changed Sex in America* (New York and London: New York University Press, 2004), 57.

[49] Tests were conducted with women who had been in their current relationships for over ten years on average and had experienced HSDD symptoms for nearly half that time. See Sprout, *Flibanserin for the Treatment of Hypoactive Sexual Desire Disorder in Premenopausal Women NDA 022526*, xv.

terms, that is, the drugs are taken following certain "lifestyle choices," for example, smoking and overeating, and drugs are taken because of issues that are more "annoyances" rather than threats to health.[50] Those definitions rely on dichotomies of legitimate and illegitimate medical conditions, blurring the lines between need and illness, and aspiration and "'legitimate' therapeutic goal."[51] The lifestyle drug that is taken for the maximisation of pleasure, such as Viagra, must nonetheless be incorporated into a medical rhetoric that serves to regulate bodies and confer legitimacy. The recognition of erectile dysfunction as a medical issue rather than an issue of sexual repression is integral to this function.[52]

The distinction between a drug of "necessity" and a drug of "recreation" is quite slippery. On Viagra and its association with "gay lifestyle," Kane Race notes:

> Viagra is coded as recreational by associating it with gay life, located squarely in the zones of leisure. This in turn works to separate out a domain of 'necessity' from a domain marked 'lifestyle,' and the distinction gets mapped onto lives and identities in culturally consequential ways.[53]

Race argues that the question of legitimate pathology and the pharmaceutical tablet is tied to market forces and capitalist interests. For him, terms such as "lifestyle" appear in the pharmaceutical domain because "they attempt to patch over the gap between medical and state or insurer determinations of what counts as necessary repair."[54] Lifestyle drugs therefore emerge in a liminal space between medical necessity and social/intimate enhancement. As chemical agents that transform bodies and practices, drugs have an inextricable social functionality. As Race writes, in their conventional prescription and consumption, drugs can only produce "a return to a putative state of normality."[55] Addyi affirms the possibility of pleasure in sexual activities all the while attempting to evacuate pleasure from its sphere. The website for instance insists that Addyi is not to be

[50] Ben Harder, "Potent Medicine: Can Viagra and Other Lifestyle Drug Save Lives?" *Society for Science & the Public* 168, no. 8 (2005): 124–125.

[51] Rod Flower, "Lifestyle Drugs: Pharmacology and the Social Agenda," *TRENDS in Pharmacological Sciences* 25, no. 4 (2004): 182.

[52] See McLaren, *Impotence*, 149–180.

[53] Race, *Pleasure Consuming Medicine*, 6.

[54] Ibid.

[55] Ibid., 7.

used for improving sexual performance. Its use is framed as therapeutic and medical.

The approval of Addyi by the FDA in August 2015 was controversial. Writers and scholars have been particularly critical of the lobbying of a conglomerate of medical associations, women's organisations and pharmaceutical corporations. The "Even the Score"[56] campaign put forward three main arguments that drew on the rhetoric of women's rights: gender equality, because men have had access to drugs for sexual enhancement for decades; the right to choose, because sexuality is a productive, necessary experience for women; and the ongoing strain that low sexual desire causes on relationships, especially within long-term, monogamous and heterosexual couples. The lobbying has been criticised as distorting feminist rhetoric to achieve financial profit, unethical manipulation of medical facts and misrepresenting the accessibility of drugs for men.[57] However, I wish to suggest here that rather than discussing whether this constituted disease mongering or a selling of sickness, we need to pay attention to the kind of subjectivity and the model of embodiment produced by the lobbying and consumption of the drug.

Scholars such as Rose and others have argued that patients cannot be viewed as passive recipients of medical knowledge or as manipulated by pharmaceutical companies. I argue, in a similar vein, that the lobbying for Addyi as well as its consumption exemplifies the emergence of a socio-technical, knowledge-gathering subject. While patients-subjects still seek medical expertise, the onus is on them to monitor and manage their appetites. And they are also expected to engage with the knowledge produced from those practices. The accessibility of information on various disorders

[56] "Even the Score" is a coalition of twenty-six organisations including Sprout Pharmaceuticals and women's health NGOs. For an analysis of this campaign, see Jacinthe Flore, "Intimate Tablets: Digital Advocacy and Post-Feminist Pharmaceuticals," *Feminist Media Studies* 19, no. 1 (2019): 3–18, and Judy Z Segal, "Sex, Drugs, and Rhetoric: The Case of Flibanserin for 'Female Sexual Dysfunction'." *Social Studies of Science* 48, no. 4 (2018): 459–482.

[57] See, for example, Judy Z. Segal, "The Rhetoric of Female Sexual Dysfunction: Faux Feminism and the FDA," *CMAJ* 187, no. 12 (2015): 915–916, and Ellen Laan and Leonore Tiefer, "'Pink Viagra': The Sham Drug Idea of the Year," *LA Times*, November 13, 2014, accessed January 19, 2019. http://www.latimes.com/opinion/op-ed/la-oe-laan-tiefer-pink-viagra-20141114-story.html.

of sexual appetite, whether through self-help books or online content, positions the patients-subjects as "knowledge-gathering" entities. There is an expectation that they will conduct research on their condition, and that they will see their physicians about potential ailments. Prior to its release to the public, the website for Addyi combined medical advice to suppliers, physicians and information for future patients and consumers. On one page, now deleted, visitors were encouraged to "Sign up for updates." The same page also featured the "Decreased Sexual Desire Screener," which visitors were advised to print for discussion with their healthcare provider. Rather than simply being sold a disease and its cure, a knowledge-gathering subject interprets and adapts the discourse of medicine into her or his narratives of the self and contributes to the clinical encounter and diagnosis. In a field of products, the subject exercises choices based on knowledge accumulated from diverse sources, including pharmaceutical providers.

The knowledge-gathering subject who seeks and consumes the sexual tablet cultivates and manages "pharmaceutical intimacy." Pharmaceutical intimacy, as a practice with material and social implications, emphasises the sociality of medicalised subjects. Instead of producing a chemical reaction that alienates individuals from natural, unfettered sexual experiences, the ingestion of pharmaceutical tablets nurtures intimacy. Emily Martin writes that drugs, through marketing and advertising, are given particular kinds of life.[58] The narrative provided by pharmaceutical corporations, alongside the biography of impoverished relations due to lowered sexual appetite, gives the sexual enhancement pill a life of necessity. It is a chemical and technical object that will impact materially on one's intimate and social relations. When the chemical tablet is combined with nomenclature, inventories, measurements and online platforms, a social subject conversant with techniques of self-improvement emerges. This form of socio-technical subjectivity yearns for durable intimate bonds but also has the technical tools to foster those bonds, whether the techniques are chemical, classificatory or therapeutic. The subject also becomes self-reliant as information can easily be accessed through online interfaces.

The practice of pharmaceutical intimacy produces a form of socio-technical subjectivity that is continually reinforced through knowledge-gathering practices. The subject is not removed from intimate contact or alienated from a "natural" experience of intimacy. Rather she or he is *more*

[58] Emily Martin, "The Pharmaceutical Person," *Biosocieties* 1, no. 3 (2006): 276.

intimate and *more* social. For Lauren Berlant and Michael Warner, dominant narratives of intimacy serve to strengthen heterosexual privilege by promoting it as an "organizing index of social membership."[59] Pharmaceutical intimacy does not disrupt this sphere; it reinforces its necessity while simultaneously mediating intimate relations through chemical reactions. It heralds then a rapport with objects as well as partners, or an intimacy where the ingestion of a chemical tablet further enhances the necessity of sexual and intimacy contact. Addyi, through a daily labour on the brain, serves to reinforce this sociality and intimacy. As Rose and Abi-Rached note,

> We, as persons, must adopt the mental states, habits, the relationship and forms of life appropriate for this work on our brains—we must shape them as they shape us. … As responsible subjects obliged to manage ourselves in the name of our own health, it seems now we have the added obligation of fulfilling our responsibilities to others by caring for our mutable, flexible, and valuable social brains.[60]

The daily intake of Addyi buttresses the responsible subject who will take care of their sexual appetite and their relationship through a chemical action on the brain. At the same time, the swallowing of the pill is not sufficient; the subject must also continually train in the social world to identify sexual encounters and learn to act on them through a continuous practice of pharmaceutical intimacy.

Future Pleasures

The development of sexual pharmaceuticals has contributed to the emergence of a subjectivity construed through an array of devices. The practice of pharmaceutical intimacy does not only involve the swallowing of the drug, as there is an ensemble of technical and social practices attached to the drug. Sexual pharmaceuticals are drugs of endurance. This means they enhance the experience of a sexual encounter, but they also suggest a horizon of hope. The narrative produced by the drugs is one of a healthier, future sex life. The "optimism" that connects "patients, practitioners,

[59] Lauren Berlant and Michael Warner, "Sex in Public," *Critical Inquiry* 24, no. 2 (1998): 555.

[60] Rose and Abi-Rached, *Neuro*, 163.

researchers and industry"[61] is one that looks to the future, towards a time "to come." That time could involve a more effective drug or, as promised by the narrative of Addyi, a healthy balanced sex life that orients them towards the future. Pharmaceutical intimacy is not always about a present pleasure. Rather its enjoyment is situated in a future time, a deferred time, which still requires practices of consumption in the present.

Thinking through the temporality of pharmaceutical intimacy offers a way to conceptualise how futurity is central to the ingestion of those drugs and how sexual appetite is itself produced *through* futurity. For Jack Halberstam reproductive heterosexual futurity follows a normative temporality that can be termed "straight time," which is structured by paradigmatic life moments such as "birth, marriage, reproduction, and death."[62] Straight time thus evokes a continuity of lineage and a perpetuation of socio-sexual norms. This temporality is always in relation to the future, as the different moments that Halberstam identifies tend towards a desirable time that is yet to come, or towards a future generation inheriting the earth. "In Western cultures," Halberstam further notes, "we chart the emergence of the adult from the dangerous and unruly period of adolescence as a desired process of maturation; and we create longevity as the most desirable future."[63]

Straight time depends on a "mature" and responsible subject who will move towards a committed and reproductive intimate partnership. The "most desirable future" is one that will contribute more directly to longevity. This future involves the endurance of certain forms of kinship, parentage and the heritage of wealth. Pharmaceutical intimacy contributes to straight time as it reinforces the resilience and longevity of the couple. The socio-technical subjectivity combines companionship, social relations and tools of sexual management. Through pharmaceutical intimacy "straight time" becomes further orientated towards futurity. Pharmaceutical intimacy organises time in a teleological manner concerned with heteronormative reproductivity.

In *No Future: Queer Theory and the Death Drive*, Lee Edelman argues that futurity is organised around a system of heterosexual reproduction, an organisation of relations that is centred on the child. The future, he writes,

[61] Marshall, "Sexual Medicine, Sexual Bodies, and the 'Pharmaceutical Imagination,'" 135.

[62] Judith (Jack) Halberstam, *In a Queer Time and Place: Transgender Bodies, Subcultural Lives* (New York: New York University Press, 2005), 2.

[63] Ibid., 4.

is "kid stuff."[64] Political movements that work to improve social order are inherently conservative since the child remains "the perpetual horizon … the fantasmatic beneficiary of every political intervention."[65] Political imaginings of a better society always incorporate the figure of child as the embodiment of futurity and continuity. They work in the name of the child as this better future always belongs to the generations yet to arrive. Hence queerness, for Edelman, as a "denial of teleology" and a rejection of heteronormative "milestones" such as marriage and reproduction, is positioned as against the child, against futurity. Edelman contends that the queer subject embodies anti-futurity, a turn against social duties of progression and continuity. For Edelman, queer identifying people should embrace this embodiment of anti-futurity and abandon the politics of hope and optimism, for the heteronormative social order has already foreclosed this possibility for them.

Reproductive futurism and domesticity are central to Edelman's arguments as the Child embodies the telos of the social order, which itself is inextricable from heterosexual kinship. The rhetoric of reproductive futurism is self-perpetuating, inviting subjects to imagine an ideal time where the world is a better place for the generations to come. But this requires them to labour in the present; it is a work that cannot be delayed though the reward certainly is. Futurity requires devoting time and resources to a project on the self in the present for a future reward or gain. It is hence no coincidence that pharmaceutical companies foreground subjects in ostensibly stable, committed heterosexual relationships—such were the relationships harnessed for the marketing of Viagra. What sexual pharmaceutical drugs promise is a future of balanced sexual reproduction. What is reproduced here is not confined to offspring. Rather, I am also referring to the continuity of sexual appetite accompanied by a regular consumption of pharmaceutical projects. Sexual appetite emerges as an object produced through ideas of improvement for the future. It is not just that pharmaceuticals promote narratives of futurity, but sexual appetite as an object of management is itself produced by and productive of discourses of futurity. Within a field of hope and optimism, the subject who devotes time to the consumption of sexual pharmaceuticals is fecund with the promise of futurity. Any form of enhancement of bodily functions and body parts is future oriented.[66] Drugs such as Addyi and Viagra contribute to reinforce

[64] Lee Edelman, *No Future: Queer Theory and the Death Drive* (Durham and London: Duke University Press, 2004), 1–31.

[65] Ibid., 3.

[66] Rose, *The Politics of Life Itself*, 20.

certain bodily capacities, although their marketing might deny this. Sexual pharmaceuticals are enmeshed in futurity. They are investments in a willingness to undertake continuous labour for a balanced future sexual self.

The futurity of sexual pharmaceuticals deploys a temporal logic of effectiveness. Viagra acts within a specific time frame and its effects are also not meant to be long lasting. The drug is taken only as needed, will usually begin to take effect within thirty to sixty minutes and "works only when you are sexually stimulated."[67] That is, the drug will not induce sexual appetite; it will merely *reveal* desire through hardness. Viagra can also be taken four hours before sexual activity and consumers are advised that this gives them "plenty of time to be spontaneous with your partner."[68] The temporal logic of Viagra revolves around immediacy as it is consumed only when desired and its action can be witnessed rapidly. However, Viagra is also associated with improved sexual performance: "a harder erection can lead to a more satisfying sexual experience."[69] While "satisfying" is not explicitly connected to duration, male sexual performance has long been associated with an ability to delay ejaculation. Understandings of erection and ejaculation produce ideas of how long an erection should be maintained prior to the release of semen. Duration is a measure of male sexual success, and this extends to a sexual performance that is consistent and reliable *over time*.[70] Viagra thus produces an understanding of temporality where sexual performance can be, first, immediately achieved, and second, constant and satisfactory across time.

The sexual pharmaceutical becomes a way to foster long-lasting relationships towards a continuity of balanced and reliable sexual appetite and performance. Viagra promises an immediate chemical reaction and suggests the possibility of consistent erections over time, while Addyi promises a configuration of time that is predicated on regular intake and delayed gratification. Addyi requires a daily ingestion of a pill before bed and does not guarantee an immediate result. The trials conducted by Sprout suggest a three to six months' intake of Addyi and minimal efficacy: an increase of around 0.8 "satisfying sexual events" per month.[71] Addyi requires a

[67] Viagra. "Learning: How Does Viagra Work?" 2019, accessed January 19, 2019. https://www.viagra.com/learning/how-does-viagra-work.

[68] Ibid.

[69] Ibid.

[70] Michael Johnson Jr., "'Just Getting Off': The Inseparability of Ejaculation and Hegemonic Masculinity," *The Journal of Men's Studies* 18, no. 3 (2010): 238–248.

[71] See Sprout, *Flibanserin for the Treatment of Hypoactive Sexual Desire Disorder in Premenopausal Women NDA 022526*, i–xvi.

daily intake at bedtime and the increase in sexual appetite is deferred to another time. Through regular chemical labour on the brain, the subject delays satisfactory sexual events to a future time. The effects of Addyi are also less noticeable than Viagra, suggesting a future occasion where the subject is better able to be receptive to a sexual encounter. Both sexual appetite and the identification of desire are suspended; they require labour in the present for a pleasure to come.

This chapter has examined how the pharmaceutical tablet has emerged in the twenty-first century as a technique for the management of sexual appetite. It has explored the creation and marketing of Addyi as a case study of how this technique produces a particular subject of pharmaceutical knowledge. Through Addyi and apparatuses of diagnosis, the subject is armed with tools to gather knowledge of sexual appetite, reflecting the development in Western psychiatry of a socio-technical and knowledge-gathering subject. The act of ingestion importantly embeds techniques within the body through chemical actions. This chapter has thus shown how pharmaceutical intimacy as a technique of self-management represents sexual subjectivity mediated by written text, spoken word and chemical interactions.

Bibliography

Addyi. 2019. *Addyi (flibanserin)*. https://addyi.com/. Accessed 19 Jan 2019.

American Psychiatric Association. 2013. *Diagnostic and Statistical Manual of Mental Disorders*. 5th ed. Washington, DC: American Psychiatric Association.

Berlant, Lauren, and Michael Warner. 1998. Sex in Public. *Critical Inquiry* 24 (2): 547–566.

Bogaert, Anthony F., and Lori A. Brotto. 2014. Object of Desire Self-Consciousness Theory. *Journal of Sex and Marital Therapy* 40 (4): 323–338.

Borsini, F., E. Giraldo, E. Monferini, G. Antonini, M. Parenti, G. Bietti, and A. Donetti. 1995. BIMT 17, a 5-HT2A Receptor Antagonist and 5-HT1A Receptor Full Agonist in Rat Cerebral Cortex. *Naunyn Schmiedebergs Archives of Pharmacology* 352 (3): 276–282.

Cacchioni, Thea. 2015. *Big Pharma, Women and the Labour of Love*. Toronto: University of Toronto Press.

Clayton, A.H., L. Dennerstein, R. Pyke, and M. Sand. 2010. Flibanserin: A Potential Treatment for Hypoactive Sexual Desire Disorder in Premenopausal Women. *Women's Health* 6 (5): 639–653.

Drucker, Donna J. 2014. *The Machines of Sex: Research Technology and the Politics of Identity, 1945–1985*. Dordrecht: Springer.

Edelman, Lee. 2004. *No Future: Queer Theory and the Death Drive*. Durham and London: Duke University Press.

Flore, Jacinthe. 2019. Intimate Tablets: Digital Advocacy and Post-Feminist Pharmaceuticals. *Feminist Media Studies* 19 (1): 3–18.

Flower, Rod. 2004. Lifestyle Drugs: Pharmacology and the Social Agenda. *TRENDS in Pharmacological Sciences* 26 (4): 182–185.

Food and Drug Administration. 2019. *Drug Trials Snapshot: Vyleesi*. https://www.fda.gov/drugs/drug-safety-and-availability/drug-trials-snapshots-vyleesi. Accessed 19 July 2019.

———. 2019. *Vyleesi Label*. https://www.accessdata.fda.gov/drugsatfda_docs/label/2019/210557s000lbl.pdf. Accessed 19 July 2019.

Foucault, Michel. 2003. *Society Must Be Defended: Lectures at the Collège de France, 1975–76*. Trans. David Macey. London: Penguin Books.

Gelez, Helene, Pierre Clement, Sandrine Compagnie, Diane Gorny, Miguel Laurin, Kelly Allers, Bernd Sommer, and Francois Giuliano. 2013. Brain Neuronal Activation Induced by Flibanserin Treatment in Female Rats. *Psychopharmacology* 230: 639–652.

Georgiadis, J.R., and M.L. Kringelbach. 2012. The Human Sexual Response Cycle: Brain Imaging Evidence Linking Sex to Other Pleasures. *Progress in Neurobiology* 98: 49–81.

Halberstam, Judith (Jack). 2005. *In a Queer Time and Place: Transgender Bodies, Subcultural Lives*. New York: New York University Press.

Haraway, Donna J. 1991. *Simians, Cyborgs and Women: The Reinvention of Nature*. New York: Routledge.

Harder, Ben. 2005. Potent Medicine: Can Viagra and Other Lifestyle Drug Save Lives? *Society for Science & the Public* 168 (8): 124–125.

Hitt, Jack. 2000. The Second Sexual Revolution. *The New York Times*, February 20. http://www.nytimes.com/2000/02/20/magazine/the-second-sexual-revolution.html?pagewanted=all. Accessed 19 Jan 2019.

Johnson, Michael, Jr. 2010. 'Just Getting Off': The Inseparability of Ejaculation and Hegemonic Masculinity. *The Journal of Men's Studies* 18 (3): 238–248.

Laan, Ellen, and Leonore Tiefer. 2014. 'Pink Viagra': The Sham Drug Idea of the Year. *LA Times*, November 13. http://www.latimes.com/opinion/op-ed/la-oe-laan-tiefer-pink-viagra-20141114-story.html. Accessed 19 Jan 2019.

Lamm, Steven, and Gerald Secor Couzens. 1998. *The Virility Solution: Everything You Need to Know About Viagra, the Potency Pill That Can Restore and Enhance Male Sexuality*. New York: Fireside Books.

Loe, Meika. 2004. *The Rise of Viagra: How the Little Blue Pill Changed Sex in America*. New York and London: New York University Press.

Lucke, Jayne. 2015. A Sexually Satisfying Event for Women, or Just a New Identity for an Old Antidepressant. *The Conversation,* June 6. https://theconversation.com/a-sexually-satisfying-event-for-women-or-just-a-new-identity-for-an-old-antidepressant-42734. Accessed 19 January 2019.

Marshall, Barbara L. 2002. 'Hard Science': Gendered Constructions of Sexual Dysfunction in the 'Viagra Age'. *Sexualities* 5 (2): 131–158.

———. 2009. Sexual Medicine, Sexual Bodies and the 'Pharmaceutical Imagination'. *Science as Culture* 18 (2): 122–149.

Martin, Emily. 2006. The Pharmaceutical Person. *Biosocieties* 1 (3): 273–287.

McLaren, Angus. 2007. *Impotence: A Cultural History.* Chicago and London: University of Chicago Press.

Moynihan, Roy. 2003. The Making of a Disease: Female Sexual Dysfunction. *British Medical Journal* 326 (7379): 45–47.

Potts, Annie. 2005. Cyborg Masculinity in the Viagra Era. *Sexualities, Evolution and Gender* 7 (1): 3–16.

Preciado, Beatriz (Paul). 2013. *Testo Junkie: Sex, Drugs, and Biopolitics in the Pharmacopornographic Era.* Trans. Bruce Benderson. New York: The Feminist Press.

Race, Kane. 2009. *Pleasure Consuming Medicine: The Queer Politics of Drugs.* Durham and London: Duke University Press.

Rose, Nikolas. 2003. Neurochemical Selves. *Society* 41 (1): 46–59.

———. 2007. Beyond Medicalisation. *The Lancet* 369: 700–702.

———. 2007. *The Politics of Life Itself: Biomedicine, Power, and Subjectivity in the Twenty-First Century.* Princeton: Princeton University Press.

Rose, Nikolas, and Joelle M. Abi-Rached. 2013. *Neuro: The New Brain Sciences and the Management of the Mind.* Princeton and New York: Princeton University Press.

Rowland, D.L., and I.G. Motofei. 2015. The Mind and Sexuality: Introduction to a Psychophysiological Perspective. *Journal of Mind and Medical Sciences* 2 (1): 1–8.

Segal, Judy Z. 2015. The Rhetoric of Female Sexual Dysfunction: Faux Feminism and the FDA. *Canadian Medical Association Journal* 187 (12): 915–916.

———. 2018. Sex, Drugs, and Rhetoric: The Case of Flibanserin for 'Female Sexual Dysfunction'. *Social Studies of Science* 48 (4): 459–482.

Sprout Pharmaceuticals. 2015. *Flibanserin for the Treatment of Hypoactive Sexual Desire Disorder in Premenopausal Women NDA 022526.* Advisory Briefing Document.

Stoléru, Serge, Véronique Fonteille, Christel Cornélis, Christian Joyal, and Virginie Moulier. 2012. Functional Neuroimaging Studies of Sexual Arousal and Orgasm in Healthy Men and Women: A Review and Meta-analysis. *Neuroscience and Biobehavioral Reviews* 36: 1481–1509.

Tiefer, Leonore. 2002. Arriving at a 'New View' of Women's Sexual Problems: Background, Theory and Activism. *Women & Therapy* 24 (1–2): 63–98.
———. 2004. *Sex Is Not a Natural Act and Other Essays*. New York: Westview Press.
———. 2006. The Viagra Phenomenon. *Sexualities* 9 (3): 273–294.
Viagra. 2019. *Learning: How Does Viagra Work?* https://www.viagra.com/learning/how-does-viagra-work. Accessed 19 Jan 2019.
Wilson, Elizabeth A. 2015. *Gut Feminism*. Durham and London: Duke University Press.

CHAPTER 7

Coda

The genealogies of sexuality written in the twentieth and twenty-first centuries have been dominated by histories of sexual identities—projects that are urgent and necessary for advancing the rights of sexual minorities, questioning assumptions that underpin understandings of sex and gender, as well as interrogating archives to shed light on our sexual present. These histories have focused on *whom* one is attracted to, rather than the dynamics of the use of pleasure. This book has instead explored what it might mean for the history of sexuality to open up a different route in its genealogy, one that foregrounds the questions of "how much?" and "how intense?" It has re-examined canonical texts in the histories of sexuality while holding on to concepts of appetite, amount, balance and frequency.

A Genealogy of Appetite in the Sexual Sciences traversed nineteenth-century Europe and journeyed through the twentieth and twenty-first centuries in the United States to explore how the problem of sexual appetite has been translated, interpreted and re-signified across time and space. The aim of this genealogy is to demonstrate how the development of psychiatric knowledge on sexual appetite in Europe in the nineteenth century, and its circulation and intensification in the United States in the twentieth century, was concomitant with the emergence of the institution of psychiatry and the medical techniques that congealed around them.

This book has argued that techniques of the patient case history, elixirs and devices, measurement, diagnostic manuals and pharmaceuticals are key to understanding how sexual appetite became medicalised in the

J. Flore, *A Genealogy of Appetite in the Sexual Sciences*,
https://doi.org/10.1007/978-3-030-39423-3_7

nineteenth to twenty-first centuries. Indeed, these techniques are deeply enmeshed in *how* psychiatry produced knowledge on sexual appetite, but also how psychiatry continually positions itself and its techniques as integral to diagnosis, management and self-care. The genealogical method employed in this book brings together different historical moments and offers not a teleological progression of knowledge but rather a genealogy of how certain questions (e.g., "how much?" and "how intense?") resurfaced and disappeared, resonated and subsided in Western psychiatry across the nineteenth to twenty-first centuries.

The terms which encapsulated the problem of sexual appetite in the nineteenth century in Europe are archaic. They included nymphomania and onanism and other more obscure terms, such as sexual hyperaesthesia and sexual anaesthesia. Twentieth and twenty-first century psychiatry, on the other hand, with assistance from the industry of sex and marriage therapy, and the APA's perennial DSM, has translated these terms into their current formulations: female sexual interest/arousal disorder, male hypoactive desire disorder and hypersexuality, amongst others. It is not incidental that this book's penultimate chapter has examined the emergence of the technique of the sexual pharmaceutical in the twenty-first century. This chapter demonstrated how the problem of sexual appetite, psychiatric treatment and practices for self-care have become understood through the enmeshment of techniques of written text, spoken word and chemical interactions. The possibilities generated by holding on to the question of appetite in genealogies of sexuality has moved us between the written text and spoken word (patient case history, the interview and its calculations, the marital sex manual, and the psychiatric manual) and the interior of the body (elixirs, physiological responses, and pharmaceuticals). Sexual appetite in the twenty-first century, however, as the sexual pharmaceutical demonstrated, necessitates the use of different psychiatric techniques (patient case history, interviews and calculations, manuals and questionnaires), *and* embodied management through the ingestion of drugs.

This book has argued that there has been a realignment of concerns with sexual disorders or dysfunction from object choice to appetite in the twenty-first century in Western psychiatry. Each chapter revealed how techniques, such as the patient case history, elixirs and devices, measurement, diagnostic manuals and pharmaceuticals were instrumental in this shift in discourses of psychiatry to the problematisation of sexual appetite. To think through appetite is to consider concepts such as amount, balance and frequency alongside object choice. While sexual appetite requires its own genealogy—and this book has endeavoured to delineate its parameters and write this genealogy—it remains germane to avoid thoroughly disentangling sexual appetite and sexual object choice. They continually inform and depart from each other, while remaining inextricable. Continuing to ask how sexual appetite circulates in the present does not require that we dispense with object choice, but rather that we interrogate how appetite manifests—sometimes in unexpected spaces—and continues to influence how we recognise and imagine ourselves as subjects of sexual knowledge.

Index[1]

[1] Note: Page numbers followed by 'n' refer to notes.

J. Flore, *A Genealogy of Appetite in the Sexual Sciences*,
https://doi.org/10.1007/978-3-030-39423-3

The manufacturer's authorised representative in the EU is Springer Nature Customer Service Centre GmbH, Europaplatz 3, 69115 Heidelberg, Germany. If you have any concerns regarding our products, please contact ProductSafety@springernature.com

Printed and bound by CPI Group (UK) Ltd, Croydon, CR0 4YY
25/09/2026
02222902-0001